AF847955

SpringerWienNewYork

Gottfried Punz

Evolution of 3G Networks

The Concept, Architecture and Realization of Mobile Networks Beyond UMTS

SpringerWienNewYork

Dr. Gottfried Punz
Stolberggasse 42B47
1050 Vienna, Austria
gottfried.p@hotmail.com

This work is subject to copyright. All rights are reserved, whether the whole or part of the material is concerned, specifically the rights of translation, reprinting, reuse of illustrations, recitation, broadcasting, reproduction on microfilm or in any other way, and storage in data banks.

Product liability: The publisher can give no guarantee for all the information contained in this book.

The use of general descriptive names, registered names, trademarks, etc. in this publication does not imply, even in the absence of a specific statement, that such names are exempt from the relevant protective laws and regulations and therefore free for general use.

© 2010 Springer-Verlag/Wien
Printed in Germany

SpringerWienNewYork is part of Springer Science+Business Media
springer.at

Credits (© Courtesy of 3GPP): p. 29, fig. 2–8; p. 75, tab. 4–1; p. 125, tab. 4–8 and 4–9; p. 274, tab. 6–6; p. 277, tab. 6–7 and 6–8; p. 278, tab. 6–9; p. 280, tab. 6–10; p. 282, tab. 6–11.

Coverdesign: WMXDesign GmbH, Heidelberg, Germany
Camera ready by the author
Data conversion: le-tex publishing services GmbH, Leipzig, Germany

Printed on acid-free and chlorine free bleached paper
SPIN: 12245840

Library of Congress Control Number: 2009940571

ISBN 978-3-211-09439-6 e-ISBN 978-3-211-09440-2
DOI 10.1007/978-3-211-09440-2
SpringerWienNewYork

Table of Contents

Chapter 1: Introduction

In this chapter we describe the motivation for writing this book and explain its scope. Some remarks on nomenclature are given in order to help the reader with a fast and easy start. The concept and structure of the material compiled is presented, followed by some hints on how to make best use of it. Finally the status of standardization, on which this book is based, is described.

1.1 Motivation

Probably the main motivation for starting to draft the script for this book, and eventually to finish it, was the desire to have a more or less complete, up-to-date overview of mobile network technology for myself, not only when starting my work in 3GPP standardization, but also continuously afterwards. I realized that some of my colleagues were in search of the same, and I extrapolated to the point where, after 3G technology is in in the field for some time, the huge, new step of development would be implemented and finally exist in reality: a manifold of system designers, SW engineers, solution consultants, test personnel, field technicians and service staff would have to deal with the underlying architecture, concepts and detailed procedures. Yet, I noticed in my roughly two decades of work as a engineer (in a few diverse fields) that compact, consistent, and balanced overview material, suitable for the wider audience is scarce. I asked myself from time to time (in the course of my daily work in 3GPP standardization), how this could be alleviated in this very case.

The second motivating factor was the feeling that, besides the daily work with its small bits and pieces of output, there is a story to tell. However, it is not found in the final result of e.g. a system design or a heape of specifications; my intention was therefore also to compile a "colourful" snapshot of the status of mobile network technology, as represented by the de facto world standard defined by 3GPP, in mid 2009.

1.2 Scope

The intention of this book is to present the complete Evolved 3GPP System of mobile networks technology, according to the specifications frozen in March 2009 (defining Release 8 - see a more detailed explanation of this term in the subsequent sub-section and in sub-section 2.4.1). Some minor features originally intended for this release, but postponed finally to the next release, are discussed in their currently available status (though further changes cannot be excluded at the time of writing).

The focus is put on the system design and core network functionality on the networking layer; it is not going into the details of the new radio layer and radio access network aspects. These constitute certainly an important sub-system of and prerequisite for the Evolved 3GPP system, but there are whole books specializing on them anyway (see e.g. [1], [2], [3]); and it is also felt that the need is more for a presentation of the evolved 3GPP system as the **umbrella** for diverse radio access technologies. Similarly, service features and functionality, even though some important ones are building blocks of 3GPP Rel. 8, are not main topics of this book; specifically, we do not discuss in detail higher layer (e.g. IMS based) services. However, we will address some impacts on higher layers arising from the base functionality of mobility and network access.

This book is also not meant as a reference guide. So, completeness is not possible and also not desirable, as it would overload the reader and not contribute to the understanding. The goal is to present the overall picture and explain how, for a complex system of an advanced mobile network, the numerous components work together, and why it is designed in that way.

1.3 Nomenclature

Various terms have been and are still used for describing the state of mobile network technology; even if all efforts were taken to adhere to a uniform nomenclature in this book, it was not seen wise to ignore other, frequently used alternative names and terms.

Evolved 3GPP System: this is the standard name used in this book for the complete description/specification of the mobile network technology, based on and successing GPRS/UMTS/HSPA and published by 3GPP by its Release 8. In particular, it includes aspects of evolution in all domains of the network (radio access network, core network and service control layer).

3G+ or 3,5G: these are more marketing terms for more advanced phases of development of mobile networks, and they are clearly not well defined. Some authors already use these terms solely with enhancements of UMTS. They are avoided in this book, although the reader can understand the Evolved 3GPP System as one of their representatives or enhancements (but is seen sometimes also as a 3,9G technology). For a good compilation and discussion of the mobile network "generations", see e.g. [4].

Long Term Evolution (LTE): this term is frequently used for the new, evolved radio access (with a focus on physical and link layer); the abbreviation

LTE is still used frequently throughout 3GPP specifications (and it is even used in the official logo on cover sheet of 3GPPs specifications, starting from Rel. 8). A term more based on technical grounds is **E-UTRA (Evolved UMTS Radio Access)**.

System Architecture Evolution (SAE): this term was used heavily for some time in the design phase of the Evolved 3GPP System, generally for the newly developped parts "above" the radio network. It was later decided that the more specific terms "Evolved Packet Core" (EPC) and E-UTRAN (denoting the architecture and functions of the whole new radio network) shall be used in specifications. The combination of EPC and E-UTRAN is termed **Evolved Packet System (EPS)**.

Release: specification of the 3GPP system is advanced in bundles of features, called "releases" (Rel.). A release is a reference set of features and it has a definite freeze date (see also sub-section 2.4.1). Compatibility between releases is considered explicitly in the design and development process, and should be considered by operators in their deployment. For this book Rel. 8 is the basis, but some hints on Rel. 9 and Rel. 10 are given as a preview. The term "Pre-rel. 8" denotes the full set of capabilities up to Rel. 7, i.e. the status before LTE/SAE.

Legacy: all what is in existence before 3GPP Rel. 8 (architecture, terminals, networks, etc); the terms "legacy" and "pre-rel. 8" can be used synonymously in the 3GPP domain. Additionally the 3G mobile network standard of 3GPP2 was seen as legacy, due to the interest and migration plans of large US operators with network deployments based on this technology (see sub-section2.4.4).

Access: in isolation, this term is used in several different meanings, related to layer 2, layer 3 or higher layers of the OSI model. E.g. we talk about "access" to a radio link/cell (as indicated by the letter "A" in terms like W-CDMA), to a Packet Data Network (PDN) and also to IMS. Generally the meaning should be clear from the context, but we noticed some confusion even in specifications, especially when it comes to combination with other terms. Often the related antonym is "core network", but for IMS the whole GPRS network is considered as "access", as indicated by "IP connectivity access network" (IP-CAN). Additionally, the term may express the procedure or an arrangement of physical/logical entities enabling a particular connectivity (e.g. "non-3GPP access"); for the the latter the more exact term "access network" is preferred.

The following two terms are used heavily in the description of architecture and functionality; although they have firm and formal definitions there is permanently slightly diiffering use of terms and interpretation:

Reference point: A conceptual point at the conjunction of two non-overlapping functional groups (source: ITU-T I.112).

Interface: The common boundary between two associated systems (source: ITU-T I.112).

We use them in a more intuitive way: in architectural graphs we use "reference point", in description of procedures realized by information flows between functional entities, we talk about an " interface".

For brevity, the symbol '@' is used for '(IP) address' in some message flows and tables, and the official name "3GPP AAA server" in many architectural graphs and message flows is shortened to "AAA server".

For simplicity, the female forms "she" and "her" are used as a pronoun for the third person (also for the "operator"); male readers should certainly not feel discriminated against.

Lastly, the pronoun "we" is used throughout the book, except in the very personal statements in this chapter, to denominate the (only one!) author. It is preferred over "I" for two reasons: for a better appearance of the printed language; and it does not confront the reader with a single person, but includes her more in the style of a joint effort (".. here we explain ..."). Certainly it is not intended as a "pluralis majestatis", but rather as "pluralis auctoris" ...

1.4 Concept of this book

It was necessary to assume some entry level of knowledge for readers of this book; although they cannot be applied too strictly, the recommended criteria are:

- medium level of knowledge of 3GPP's 2G/3G circuit and packet switching technologies, i.e. GSM and GPRS architecture and procedures, preferably also 3GPP's Interworked WLAN (I-WLAN). It is expected that the reader is already familiar with terms used heavily in 3GPP's 2G/3G system (like "PLMN", "handover", "roaming", "paging", "circuit switching (CS)" / "packet switching (PS)", "User Equipment (UE)", "frequency band" etc.);
- familiarity with the OSI layer model and related protocol usage;
- basic knowledge of layers 1 and 2 of at least one radio access technology (e.g. GERAN/UTRAN, WLAN i.e. IEEE 802.11, WIMAX i.e. IEEE 802.16);
- knowledge of the Internet Protocol on medium level, especially addressing and networking for IPv4 and IPv6, and features related to mobility, authentication/authorization/accounting (AAA) and Quality-of-Service (QoS);
- the notion of control plane and user plane should be understood (it is well defined in 3GPP technology, but less clearly defined in IP technology);
- some basic knowledge of 3GPP's Internet & Multimedia subsystem (IMS).

There are various sources available for the reader to acquire this prerequisite knowledge, e.g. see [5] and [6].

The complexity of the topic requires a phased approach, similar to the phases stage 1 to 3 used in standardization. Stage 1 (requirements) map onto chapter 2, where reasons and needs for evolution of the 3GPP system are presented. Although we will not aim at listing requirements in their completeness (this would look more like a design or development document), the main drivers are discussed and explained. Stage 2 (architectural design) is mainly covered in chapter 3, but additionally chapter 4 gives more detailed explanations on the underlying concepts. A reader may therefore consult quickly chapter 3, to see the overall picture and to grasp a specific architectural aspect and follow it selectively for its details in subsequent chapters. Alternatively, she can read chapter 3 and 4 in full, in which case she will acquire a solid and broad understanding of the Evolved 3GPP system on architectural and functional level; it then depends on whether she likes to go through the details in the same broadness, or postpones that for later. Chapter 5 and 6 dip progressively into these details, first on procedural and then on protocol level. It is also recommended to return to previous chapter(s) every now and then, in order to compare the presentations on these different levels and consolidate the view.

This book does not aim at the expert level; a specialist may find explanations and details too coarse, from her point of view. Yet, this is indeed the intention, not to overload the reader with the total amount of information, but provide her with the oversight and principal understanding in one place, where she would have to read dozens of lengthy specifications (which is naturally a tedious task). These specifications are extremely condensed, containing exact wording and including all small bits, but probably not the desired explanation for the non-specialist. And we believe that also an expert within one the diverse fields contributing to this technology, who is normally not able to follow all streams of development, would benefit from the broader view.

We have simplified the message flows reasonably and limited to the "good" case, the numerous error cases are left out either completely, or only a short hint is given in the descriptive text.

References are numbered and listed per main chapter.

1.5 Status of standardization at the time of writing

Standardization of mobile networks, in particular within 3GPP, progresses steadily; at any time it is thus only possible to present the intermediate status. With the next release enhancements, and in some cases removal, of functionality will be done.

At the time of writing (July 2009) the situation was as follows: requirements for 3GPP Release 8 (i.e. stage 1 of the system specification) had been already frozen in December 2007 (with a few exceptions till June 2008) and system/security architecture (stage 2) in June 2008 (again with a few exceptions till September 2008). In its December 2008 plenary meetings, 3GPP decided to freeze the specifications in principle also for protocol and other detailed functionality (i.e. for stage 3), but with an appreciable amount of exceptions for 3 months. These exceptions were fulfilled in the one remaining meeting cycle and in March 2009 3GPP declared the whole Rel. 8 as completed.

This does not mean that Rel. 8 specifications are not touched any more; it is quite natural for a system of such complexity that unclarities and errors are detected, especially when massive implementation then starts. It means that errors need to be corrected in the specifications by so-called category "F" change requests. Their number is still high at the time of writing, and it is quite likely that some text on details (of message parameters and procedures) written down at best knowledge before is no longer correct. However, it is not likely that something in the presented material changes fundamentally.

As a matter of fact, before LTE/SAE could be finalized fully, the next step of development already shaped up with the so-called "LTE Advanced". At the time of writing the technical study work was on the way within 3GPP's RAN groups, accompanied by high-level coordination (e.g. submission to ITU-R, proposing it as a candidate for IMT-Advanced).

1.6 Acknowledgments

I wish to express my thanks to my colleagues in the Network Research Division of NEC Laboratories Europe/Heidelberg, and in 3GPP SA2, CT1 and CT4 workgroups, for all the valuable discussions we had during the past few years, for their sharing of explanations and opinions in many topics; such an interactive and supporting environment was indispensable for me during writing.

For their professional review I am grateful to Stefan Geiger (Siemens Austria/Vienna), Tarik Taleb and Hans-Jörg Kolbe (both NEC Laboratories Europe/Heidelberg). Despite their great effort, all remaining errors and inconveniences of text and figures are clearly to blame on me. The reader is encouraged to report them, together with any further suggestion for improvement, to:

gottfried.p@hotmail.com

We will try to include them to the extent and at the earliest point in time possible in future editions.

References

[1] Dahlman, Parkvall, Skold and Beming: "3G Evolution: HSPA and LTE for Mobile Broadband", Academic Press, Oxford, UK (2007)
[2] H. Holma, A.Toskala (Editors): „LTE for UMTS - OFDMA and SC-FDMA Based Radio Access", John Wiley & Sons (2007)
[3] S. Sesia, M. Baker, I. Toufik (Editors): "LTE: The UMTS Long Term Evolution: From Theory to Practice", John Wiley & Sons (2009)
[4] 3G Americas' White Paper (compiled by Rysavy Research): "HSPA to LTE-Advanced: 3GPP Broadband Evolution to IMT-Advanced (4G)" (September 2009)
[5] T. Halonen, J. Romero, J. Melero (Editors): "GSM, GPRS and EDGE Performance: Evolution Towards 3G/UMTS", John Wiley & Sons (2003)
[6] M. Poikselkä, G. Mayer, H. Khartabil, A. Niemi: „The IMS (IP Multimedia Concepts and Services in the Mobile Domain)", John Wiley & Sons (2009)

Chapter 2: Mobile Networks Evolution

In the following part we discuss why the next, large step of development was needed in the middle of the first decade after the millenium, at a time when the 3rd generation (UMTS, and its optimizations like HSPA) had not yet taken on full speed, and the bulk of mobile communication was still voice. We state the the driving factors and requirements formulated for an evolution, and list the limitations encountered with the legacy systems. In order to provide the reader with the baseline of the evolution we present condensed views of the main tracks in standardization of mobile and fixed networks. The chapter is concluded with snapshots on standardization history and operator commitments for future deployment of the Evolved 3GPP system.

2.1 Success and limitations of GSM, GPRS and UMTS

The success story of GSM (2G) is unprecedented and well known by own experience to virtually 100% of the population in industrialized countries, and an increasing portion of that in developing countries. This technology (see e.g. [1] for a compact overview) provides for speech (=narrowband, real-time, circuit-switched) communication in large areas, for users with mobility demands from stationary (yet wireless) to fast–moving and worldwide roaming. In 2009 many hundreds of GSM networks were in operation. Extensions were made in GSM to enable data communication (WAP, HSCSD), but the success was very limited.

GPRS brought along for mobile users the connectivity to packed data networks, in practice to IP networks including (actually almost exclusively, Internet). The data rate in the base version went up to 172 Kb/s in theory, but operators hardly ever allocated the maximum of the then necessary 8 logical channels for one user. A major blocking point were for some time the high charges, which are still excessive in roaming as of today. GPRS has the concept of a 2 stage access to IP connectivity: the first step is to attach to and register with the network, but transmission of user data requires the establishment of a so-called "PDP (Packet Data Protocol) context". Only at this point the IP address is assigned, together with other properties of this "packet bearer" (notably the actual point of interconnection to a Packet Data Network and QoS). In this respect GPRS conceptually does not realize the "always-on" IP connectivity, although the lifetime of a PDP context in practice can be stretched to days. In the popular terminology of "generations" of mobile network technologies, GPRS can be seen roughly as 2,5G. For a presentation of GPRS technology, see e.g. [2].

For both GSM/CS and GPRS/PS incremental optimizations were made, based on higher modulation efficiency under the term "EDGE" (Enhanced Data Rates for GSM Evolution), but these technologies did not change anything fundamentally.

The next, genuine 3G generation of mobile networks, UMTS (for a compact description the reader may e.g. consult [3]) , built on a new radio technology (Wideband CDMA) and promised at least two things: for the operator more bandwidth due to new radio spectrum, and higher peak data rates for the end user. The latter was announced to be theoretically 2 Mb/s under optimal conditions (stationary, in cell center), but was considerably lower in practice. The UMTS network architecture kept the parallelism of CS and PS. For the latter, a whole service control layer was created in form of the IMS (Internet and Multimedia Subsystem).

UMTS was again improved incrementally for higher data rates by HSPA (divided into downlink/HSDPA and uplink/HSUPA) and HSPA+, based on even more sophisticated radio schemes. HSDPA was standardized with 3GPP Rel. 5, HSUPA in Rel. 6 and HSPA+ in Rel. 7. This path was even continued up to Rel. 9, complementing the LTE/SAE work.

The PS core network of GPRS and UMTS (i.e. not counting IMS as the service control layer) does not foresee a split of control and user plane by separate, dedicated network nodes. (There was, however, such a split defined in the CS domain, with a functional decomposition of the main CS core network node, the MSC, into MSC server and CS media gateway; for a basic view of the corresponding architecture see sub-section 2.4.1.)

The benefit of high security and unique subscriber identification within 3GPP's network was always recognized; however, there is a difference in strength of security between the 2G and 3G networks. For the new system to be developed only the highest level of security could be allowed.

A detailed analysis cannot be performed here, but already a quick look into the architecture of the legacy network in sub-section 2.4.1 reveals that it suffers from the many node types and levels of hierarchy present. (Note: an evaluation or comparison should always be made for the same resulting level of functionality; i.e. when analyzing the new 3GPP system in this respect the additional functionality/nodes for integration of non-3GPP access networks should be taken into account). An incremental improvement was achieved already within the legacy PS technology by the so-called "Direct Tunnel" approach (it consists in bypassing one network node in the user plane under favourable conditions, see sub-section 3.4.2). But it was clear that a deeper architectural change was needed to achieve this goal.

Another area of improvement in the legacy technology can be identified with spectral efficiency, i.e. the effective number of bits deliverable per radio frequency unit and time unit. Even though new radio spectrum has been made available for mobile communication, the pressure for cost reduction and competitiveness required further gains.

2.2 The need for a new architecture

2.2.1 General

The overall work for a new 3GPP system was started with a workshop for 3GPP's RAN (Radio Access Network) standardization groups in November 2004 in Toronto, where operators, manufacturers and research institutes presented their proposals, and the first set of requirements was compiled. It was soon detected that a similar strong demand resulted for the evolvution of the mobile core network. 3GPP then performed studies for the evolution of UTRA/UTRAN, and the packet core network, before starting the actual design of the evolved system. These studies were concluded in June 2006 and December 2006, and documented in 3GPP TS 25.913 [4] and 3GPP TS 23.882 [5], respectively.

The most relevant factors for competitiveness over more than a decade were seen in:

- reduced latency for both control plane and user plane,
- higher user data rates at higher spectral efficiency,
- more flexibility in radio spectrum utilization,
- improved system capacity and coverage,
- reasonable terminal complexity and power consumption, and
- reduced cost (CAPEX and OPEX) for the operator, especially also measured per bit.

At this point a few statements on mobile network technology design can be stated. The three most important targets therein are to provide mobility, security and Quality of Service (QoS) . These are fully interlinked, as sketched in Figure 2-1 and discussed below.

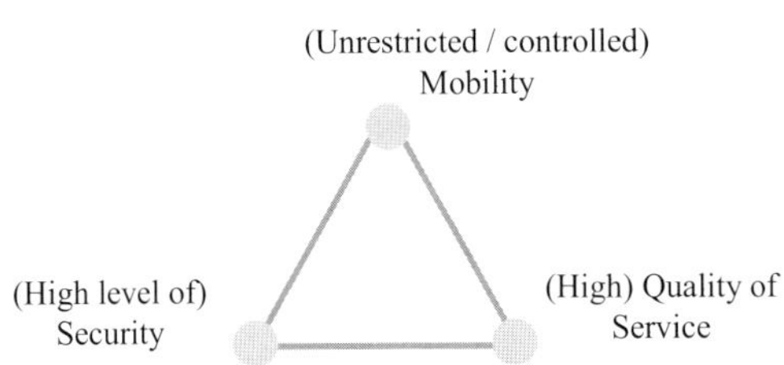

Figure 2-1: interrelation between targets in mobile network design

Mobility ←→ Security: if a user moves around and her terminal therefore needs to access the (core) 3GPP system repeatedly via different access networks, the authentication most likely has to be repeated, as in this case it is not within the same security domain. This is generally the case with non-3GPP access networks. On the other hand, realizing such a security domain for the 3GPP access domains requires to foresee exchange of context information, negotiate between involved network nodes, and select one out of the variety of supported security algorithms, as they have evolved over time.

Mobility ←→ Quality of Service: with mobility, the data transmission across the air interface is inevitably affected. First of all, different radio access technologies with varying data rates and delays may be involved; secondly, the resource utilization status in these accesses may differ even if they belong to one and the same operator, and it is not directly controllable between operators (roaming agreements would only constitute some rough and very static level of control). Finally, mobility per se, i.e. handovers, most likely disturb the normal flow of data (potentially causing delay, jitter and loss of packets); depending on the capabilities and technology of the involved access networks, type of handover and the effort spent for implementing the concrete handover procedure, these effects may be negligible, noticeable but tolerable by the user, or lead to unacceptable service degradation.

Security ←→ Quality of Service: QoS is the real value of mobile operators' service delivery, but also contributes considerably to the costs; as such, illegal use or upgrading of resources must be prevented. A sophisticated security concept is indispensable. With the shift towards the all IP paradigm, the operator's traditionally walled network environment is opened up more and more. Seen in a wider context of QoS perception, scenarios of unsolicited communication additionally must be taken into consideration. Several standardization fora (IETF, TISPAN and 3GPP) have studied this issue and concluded that special means are required for prevention.

It will be shown in the subsequent chapters how these three main targets are fulfilled to a higher degree in the evolved 3GPP system, compared to the legacy (2G and 3G) systems.

2.2.2 Radio interface and radio network aspects

The amount of wide area data transmission by mobile users had picked up only gradually after introduction of UMTS in general and UTRA (Universal Terrestrial Radio Access, the radio technology of UMTS) in particular. But for local wireless transmission (e.g. WLAN) and fixed line access (as per xDSL) it increased at a much higher rate, thanks to the free radio band in the WLAN scenario and the flat rate charging paradigm in xDSL case. Data transmission rates equal to the category of fixed line broadband were therefore considered necessary for the new system. The then upcoming radio and networking technology of WIMAX already set high targets for transmission rates. It was clear that the new 3GPP radio technology Evolved–UTRA (E-UTRA, synonymous with the LTE radio interface) had to become strongly competitive in this respect. Thus the following target transmission rates were defined (as peak data rates, i.e. not for extended times and in principle to be shared by users in a cell):

- Downlink: 100 Mb/s
- Uplink: 50 Mb/s

These numbers are valid for a reference configuration of two antennas for receptions and one transmit antenna in the terminal, and within a 20 MHz spectrum allocation.

Regarding the radio spectrum allocation, a more flexible scheme was proposed, allowing on the one hand smaller and also wider bands than in UTRA (where a fixed band width of 5 MHz is used). The supported radio band widths are 1.25 MHz, 1.6MHz, 2.5 MHz, 5 MHz, 10 MHz, 15 MHz and 20 MHz bands. At the same time the spectral efficieny, i.e. the number of bits transferred per second per Hertz, was required to be more than doubled (compared to UTRA).

The target for the E-UTRA was also to increase relatively the bitrate at the cell edge (although from radio transmission principles the decrease of effective data rate at the cell edge is unavoidable). Compared to the 3GPP technology baseline at the time of stating the requirement at least a factor should be gained.

From deployment and site acquisition and maintenance point an important target was to be able to maintain the same physical sites (i.e. locations of radio base stations/antennas); it would just be too expensive for an operator to add another set of many thousand sites for introducing the new radio technology; of course this does not exclude that for capacity increase in high density areas additional sites are used).

The latency requirements for E-UTRAN (which is constituted by the network of all E-UTRA base stations) were described by a user-plane setup of less than 10 ms between the UE and the first user plane above the radio node (which became in the final architecture the Serving Gateway). Also the control plane latency was required to decrease significantly; the transition time from the state "camped on a cell" to transmission of data was targetted at less than 100 ms. Note that these numbers (1) exclude downlink paging, and (2) constitute a requirement shared with the core network.

The E-UTRAN as a radio access network was radically simplified to one node type only, and a minimum of interfaces (see the architecture including comparison with the legacy case in sub-section 3.2).

The mobility with quality of service is represented by the handover performance; the intra E-UTRAN handovers should perform better than those in the legacy system. Building on the remarkable successful handover concepts developed for the legacy radio technology, the goal here was to provide flexibility to operators and their service delivery, while minimizing the complexity. Therefore seamless and lossless handovers were to be supported: a seamless handover aims at minimization of delay (which is critical for some services like voice), but does not guarantee the delivery of every data packet during the handover; in this case means for retransmission can be provided by upper protocol layers. A lossless handover involves sequence numbering of data packets and a tight control signaling during the handover.

2.2.3 All IP vision

A very general principle was set forth for the Evolved 3GPP system; it should be "all IP", meaning that the IP connectivity is the basic bearer service provided to the users. All upper layer services like voice, video, messaging, presence etc. are built on that. The "always on" or "always IP connected" feature is just another view of the same principle: the IP connectivity should be available immediately after registration with the network and not only on specific demand. The IP version strongly favoured was IPv6 (due to its huge address space, which was thought to alleviate the IPv4 address exhaustion timeline). But the pressure to support also the large existing base of IPv4 devices/servers/networks over a long time was encountered, and so the Evolved 3GPP system itself is, by its design concept, dual stack (i.e. suitable for IPv4 only, IPv6 only and IPv4 and IPv6 capable devices). At the time of writing, large scale operation of IPv6 based infrastructure is not yet in place; on the contrary, 3GPP agreed to study in more detail the IP version migrational scenario. An additional aspect was that for IPv4, the use of private IP addresses in corporate PDNs should not be restricted.

By dipping into the details of the architecture and looking at the protocol stacks for interfaces between network nodes, it will become clear that the "simple" model of IP (seen as the third layer in the ISO/OSI model of open systems interconnection), is not applicable to a mobile network. There are "virtual layers" plugged in between, in the form of "tunnels", realizing the three key aspects explained above: mobility, security and quality of service. As a result, IP based protocols appear both on the transport layer (between network nodes) and on higher layers. The radio interface is anway not based on pure IP protocol but on specialized radio protocols.

During development it was noted that the legacy CS part of mobile networks has still such a large installed and persistent base that a rollout of the evolved system by operators would occur only gradually, starting from local areas (e.g. densely populated metro areas). Due to this fact, some schemes for seamless interoperation were strongly requested by operators and also standardized (see the dedicated sub-sections 4.14.2 for Single Radio Voice Call Continuity (SRVCC) and 4.14.3 for CS Fallback (CSFB).

2.2.4 Diverse deployment and operation

The new system was required to be equally suited for the standalone deployment scenario, as well as an integrated deployment with legacy 3GPP systems (based on radio accesses GERAN and/or UTRAN, as well as corresponding core networks).The mobility of users was also considered to vary widely, from stationary to high velocity.

All kinds of data traffic types were required to be handled efficiently by the new system: real-time and non-real-time, high and low volume, steady and intermittent – reflecting services like voice and video conversation, streaming, messaging, metering and control. This should care for upcoming usage scenarios like machine to machine communication, ambient intelligence, etc. The differentiation of services, like massive, bit-pipe data transfer on the one hand and highly valued services (e.g. utilizing location information or best voice continuity) should always be possible for the operator, both by policing them (allowing/restricting them) and charging them differently.

The rollout, operation and maintenance of the new system should also bring along self-configuring, self-optimizing and self-healing features, for the sake of reduced OPEX. It is simply mandated by the increasing number and complexity of network nodes (most notably, radio base stations) and configuration parameters, and the need for running parallel infrastructures. The main focus here is initially the self-establishment of radio base stations, including the set up of their neighbour lists and their IP connectivity.

The need for more flexibility in radio band usage was already discussed in subsection 2.2.2; this is especially important for re-farming of radio bands, as new spectrum also becomes available. Tthe operators should have a wider choice for their overall deployment strategy and regarding their system rollout, coverage, support of features and migration along the 3GPP releases

2.2.5 Umbrella system view

The legacy 2G mobile network techology, and to a lesser extent the 3G technology, offers virtually 100% coverage and wide area connectivity with extensive roaming capability. The Evolved 3GPP system was therefore deemed ideally suited to become the umbrella for other radio technologies, which are designed more for the local coverage,e.g. hotspots and metro areas. It was desired to support the integration of their radio access networks with the core network of the Evolved 3GPP system.

Figure 2-2 illustrates a potential coverage scenario (as seen by one 3GPP operator, i.e. no roaming aspect included), together with a potential "trajectory" of a user shown by a dotted curve.The largest coverage is assumed to be provided by the legacy 2G/3G access; the LTE coverage is included therein, and is well coordinated with the legacy access (this is important for intra-3GPP handovers and for fallback scenarios). Further, appreciably smaller coverage "bubbles" are seen, and in general the must be assumed to be uncoordinated with the 3GPP network deployment. A UE, moving along with the user, crosses many times the boundaries of these coverages areas.

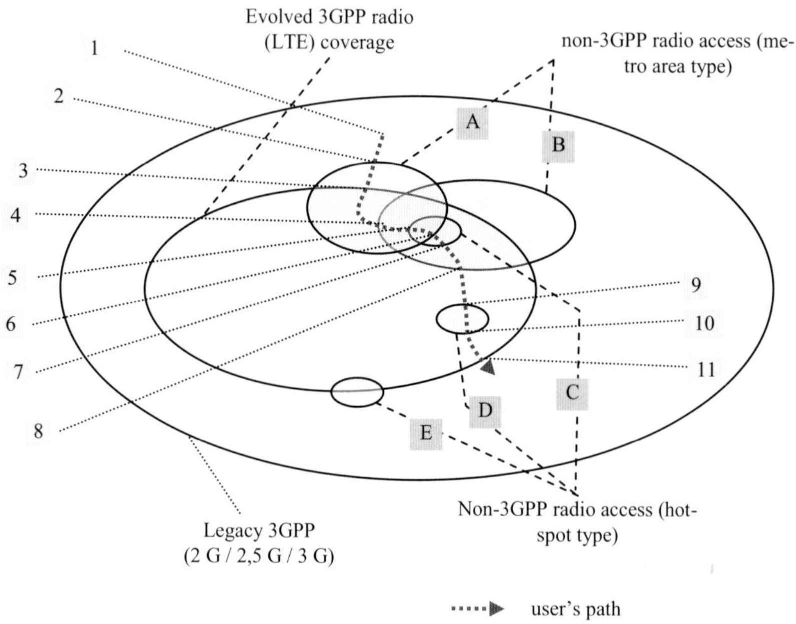

Figure 2-2: example coverage scenario for the evolved 3GPP system

The resulting decision points for access network selection and/or handover (depending on whether idle or active mode is applicable at these points) are listed in Table 2-1. The assumption here is that the non-3GPP accesses of type "metro area" have a trust relationship with the 3GPP operator, but not those of type "hotspot". It is also plausible, due to the more localized nature of both of them, that their multiplicity is much higher than that of full-size 3GPP networks.

Point	Choice between access systems	Comment
1	none	only legacy 3GPP available
2	legacy 3GPP or non-3GPP access	trusted non-3GPP access A becomes available
3	legacy 3GPP or trusted non-3GPP access A or LTE	LTE access becomes additionally available
4	legacy 3GPP or trusted non-3GPP access A or trusted non-3GPP access B or LTE	trusted non-3GPP access B becomes available

5	legacy 3GPP or trusted non-3GPP A or trusted non-3GPP B or LTE or untrusted non-3GPP access C	untrusted non-3GPP access C becomes available
6	legacy 3GPP or trusted non-3GPP access B or LTE or untrusted non-3GPP access C	trusted non-3GPP access A is no longer available
7	legacy 3GPP or trusted non-3GPP access B or LTE	untrusted non-3GPP access C is no longer available
8	legacy 3GPP or LTE	trusted non-3GPP access B is no longer available
9	legacy 3GPP or LTE or untrusted non-3GPP access D	untrusted non-3GPP access D becomes available
10	legacy 3GPP or LTE	trusted non-3GPP access D is no longer available
11	none	LTE is no longer available

Table 2-1: decision points for example coverage scenario and UE trajectory

It becomes immediately clear that the optimal discovery of and selection between access networks, as well as efficient handovers between them, is of utmost importance for both the service experience of users and the service offering of the mobile network operator. QoS, charging and configuration aspects are obviously challenging all cases except the most simple ones (e.g. considering more than one 3GPP operator expands the list of decision points).

2.3 Requirements for the new architecture

In addition to radio interface specific requirements and general boundary/preconditions discussed so far, 3GPP collected determining requirements for the new Evolved Packet Core in their specification 3GPP TS 22.278 [6]. Fulfilling these, the new 3GPP system also meets the requirements for the NGMN, as set forth in [7].

2.3.1 General targets and capabilities

In addition to the features discussed so far, the following capabilities required for the new 3GPP system are to be mentioned:

- good scalability, separately for user plane and control plane (note that this does not automatically mean separate nodes for the two planes, as it could equally be achieved with nodes combining user and control plane by appropriate traffic mixing);
- support for terminals requiring different type of mobility: fixed, nomadic and mobile terminals;

- minimal transport and signaling overhead, especially over the air;
- the signaling in idle mode for dual mode UEs (capable of the new and legacy 3GPP radio access) should be minimized;
- multicast capability on the radio channel;
- re-use or extension of well established concepts like roaming restrictions and network sharing;
- compatibility with legacy roaming principles (looking into the details of mobility protocols, this may only be true for the legacy mobility protocol GTP, but not for the new mobility protocol PMIP; see the corresponding discussion in sub-section 3.3.1);
- quite naturally, the request was for lower latency in user plane and control plane (corresponding to communication delay and communication setup delay) than in the legacy 3GPP network:
- the maximum transmission delay was required to be comparable to fixed network, concretely less than 5 ms;
- the target for control-plane latency (corresponding to a transition from idle to fully active state in the core network) was set to less than 200 ms (and reaches actually the 100 ms);
- the system complexity was to be reduced: this must be seen in relation to the functionality provided. Looking at the Evolved 3GPP system in full, it may not seem less complex than the legacy 3GPP system, but that is due to the immense increase in functionality;
- another strong desire was to arrive at a flatter architecture. Looking into architectural graphs in chapter 3, it seems that this goal has not been achieved fully in the core network; and indeed, the pressure for even more "flatness" has been revived strongly just recently (refer to the discussion of Local IP Access in sub-section 3.10.2);
- as already mentioned briefly, reduction in CAPEX/OPEX (Capital and Operational Expenditure) for operators: the assumption was that much of the potential cost saving could be achieved by relying more on general standards, like IP technology and open interfaces;
- the strong control features for operators in the 3GPP architecture should remain also with the new 3GPP system. As an example, there are numerous places in the specifications where an "operator policy" is mentioned;
- seamless operation of both real-time (e.g. VoIP) and non real-time applications and services was requested, so that packet loss and interruption times due to mobility are minimized. The system should perform well for the VoIP service. Special attention was also paid to seamless continuity for voice communication with legacy systems (both 3GPP and 3GPP2);

- support of (home operator controled) local breakout of traffic in a visited network. The feature relates to the situation that a local PDN connection (in the simplest case, internet access) can be provided for a roaming user. Clearly, this helps to achieve optimize routing of data packets, as local traffic does not need to be routed via the home operator network. On the other hand it must be under the control of the home network operator whether she wishes to handle (and potentially inspect) this traffic or not.

The efficient support of an broadcast mode for services like Multimedia Broad- and Multicast (MBMS) was required, similar to the legacy system; an increasing importance of these services was anticipated (e.g. for mobile TV). However, although the radio part was designed within Rel. 8, the time for specification of all system aspects was too short, and so this will be completed only in Rel. 9.

2.3.2 Multi access and seamless mobility

A central requirement was stated with respect to the increasingly diverse environment expected for users and terminals in the future. Access to the new system and mobility should be supported not only for the legacy and the newly to be defined 3GPP radio accesses, but also for devices utilizing existing wireless and wireline technologies (e.g. WLAN, WIMAX and fixed line broadband), as well as for emerging ones.

With mobility between different access systems, the QoS was required to be retained end-to-end (assuming available appropriate resources in the access network), and appropriate mappings to be done by the system; if a target access system does not provide the same QoS, it was required to ensure at least the service continuity.

In the legacy system, seamless handovers were in practice often limited to mobility within one network (we use the established term "Public Land Mobile Network"/PLMN from now on). Inter-PLMN handover, i.e. handovers between two PLMNs) is a definite requirement at least between E-UTRAN accesses and for PLMNs of the same EPC protocol type; until the general interworking between GTP and PMIP protocols is solved, a restriction applies here in practice for EPCs of different protocol type.

The mobility requirement extends up to speeds of 350 km/h (necessary for high speed trains). Higher speed, up to 500 km/h, can be supported in special frequency bands.

Considering the multi-access aspect, a need was seen to inform the mobile terminals (and users) on available radio accesses/ access networks and their properties. For the operator it is important to be able to control the access utilized by users flexibly (see also the discussion of coverages in sub-section 2.2.5). Yet, the capability for extended parallel operation via several radio interfaces/accesses was not yet seen in scope (this is subject to further work in Rel. 9).

2.3.3 Security and privacy

A general principle set forth regarding security of the new system was, that it must provide an equivalent or higher level of security than the Rel. 7 3GPP system. Also, a breach in security in one access technology must not impact the security in other accesses. As an all-IP system, the special threats due to Internet connectivity must be countered. A (3G) USIM application on UICC is required for authenticating and authorizing a user for EPS services, i.e. a (2G) SIM is not enough. As in the legacy system, lawful interception of user traffic and signaling needs to be supported and corresponding data provided to authorities.

Special care had to be taken for the mobility management; e.g. unsecure mobility signaling could lead to denial of service attacks. Location hiding was also required, configurable and controllable per user by the operator; this effectively may also result in the denial of certain traffic route optimization (e.g. a local breakout in the visited network).

User identity confidentiality consists in use of temporary identifiers, as much as it is possible. However, there remain still cases were the permanent identity is carried over the radio link.

2.3.4 A special feature: emergency warning

The 2004 Tsunami disaster in South East Asia gave rise to a a strong regulatory push, mainly from Asian authorities, for a warning support through mobile network infrastructure. Requirements were formulated in an extremely tight manner, namely to be able to alert mobile users with a primary notification within 4 seconds (counted from the delivery of a triggering request to the mobile network operator), and subsequent transmission of secondary warning information. In Rel. 8 the basic Earthquake and Tsunami Warning System (ETWS) was to be standardized, so that such warnings could be delivered in an efficient way, confined to the affected area and protected (against spoofing, which could eventually cause major uproar and damage). ETWS targets both at the LTE/EPC and legacy systems.

Note: in 3GPP Rel. 9 a more general and global public warning system will be defined for the global market, including the commercial mobile alert system foreseen for the US. The emergency events are described by the affected area, recommended action, expiration time (with time zone) and sending agency.

General emergency communication, i.e. the counterpart of legacy emergency calls (by Voice-over-IP service via E-UTRAN access and IMS) was not seen in scope of Rel. 8; it will be defined as one of the building blocks of Rel. 9.

2.4 Evolution tracks in mobile network standardization

2.4.1 3GPP (3rd Generation Partnership Project)

3GPP is the most prominent standardization group for mobile networks and in existence since 1998; it therefore has well established, but also rather strict procedures. Their specifications come in bundles called "Releases", see Table 2-2; originally the plan was to have yearly releases, but this was relaxed, and releases are now planned more flexibly. The amount of 3GPP specifications is on the order of several tens of thousands of pages for Release 7 (the last one before LTE/SAE). 3GPP inherited also the maintenance of GSM and GPRS specifications. 3GPP's organizational structure and resources are huge and funneled by all large vendors and operators; so is their impact.

Release	**Published**	**Key architectural features**
Rel. 99	March 2000	UTRAN (RNC, NodeB), USIM
Rel. 4	March 2001	MSC split into MSC server and Media GW
Rel. 5	March 2002	IMS, HSDPA, IP based UTRAN
Rel. 6	March 2005	I-WLAN, HSUPA, MBMS, Service Based Local Policy, Flow based charging, IMS enhancements (PoC)
Rel. 7	December 2007	Unified PCC, Direct Tunnel (of user plane between RNC and GGSN), MIMO,HSPA+, Common IMS, VCC

Table 2-2: 3GPP releases (from Rel. 99 to Rel. 7)

The requirements for the evolution of the 3GPP system specifically also addressed the integration of other technologies (as access systems) and the interworking with and migration from the legacy 3GPP system. In this respect 3GPP has the widest scope.

For reference and comparison the legacy 3GPP architecture (2G, 2,5G, 3G) is given in Figure 2-3 in a simplified form (e.g. SMS entities, charging interfaces and interfaces used for node relocation are not shown, and no aspects of roaming/PLMN interconnection are present). It was also the starting point for the design for the new system, and the subject of integration. Reference points are indicated for completeness and illustration (the interested reader is asked to consult the official 3GPP Network Architecture [8]).

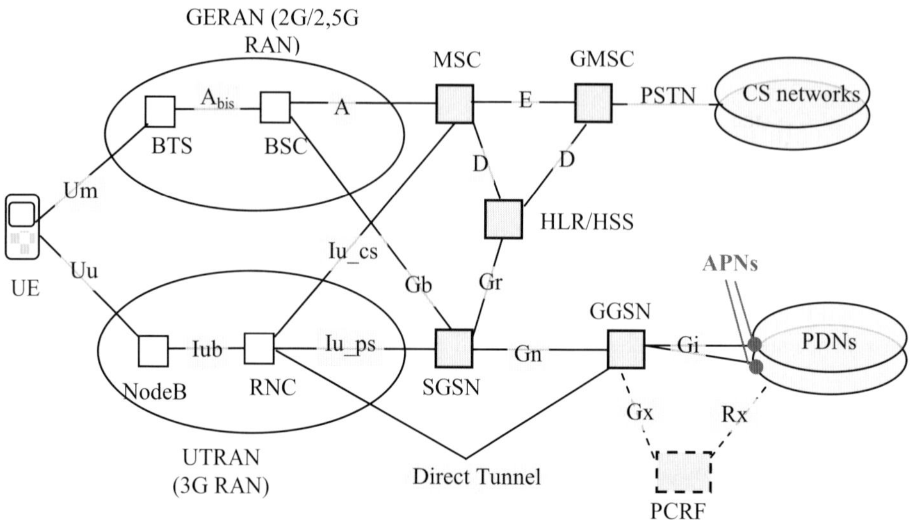

Figure 2-3: legacy 3GPP architecture (for 2G, 2,5G and 3G accesses; simplified)

The network entities in brief are:

UE	User Equipment: the mobile terminal
BTS	Base Transceiver Station: the 2G/2,5G radio base station
BSC	Base Station Controller: a controlling node in the 2G radio network
NodeB	3G radio base station
RNC	Radio NW controller: controlling and concentratingnode in the 3G radio network
(G)MSC	(Gateway) Mobile Switching Center: circuit switched core network nodes
S/GGSN	Serving/Gateway GPRS Support Node: packet switched core network nodes
HLR/HSS	Home Location Register / Home Subscription Server: central data base
PCRF	Policy and Charging Rules Function: a control node for policy managemant and charging (optional)

The last four of them (marked in grey in the graph) are representatives of an mobile operator's core network; on their left we find the radio access networks. Reference points are included for completeness, but only a few of them will be referenced subsequently (Gn, Gp, Gx, Rx). The Internet and Multimedia Subsystem (IMS) is not shown as a separate entity (it can be see as a PDN).

The concept of Access Point Name (APN) is explained briefly here and visualized in Figure 2-3, as it is used analogously in the Evolved 3GPP system: an APN defines the point of interconnection with the PDN; it identifies the edge node for the connection to the PDN (in GPRS it is the GGSN), but moreover also the corresponding connection point in the PDN. This is necessary, because one and the same GGSN may support connections to different PDNs. Additionally an APN identifies also a service, the simplest one being pure IP connectivity; an example for a more elaborated service would be IMS, and the corresponding APN would then address the/a P-CSCF. APNs need to be authorized for use and are linked to overall QoS parameters (e.g. a maximum bit rate). The detailed format and coding of APNs is given in sub-section 4.21.6.

HSPA, an optimization of the radio transmission, was introduced in 3GPP Rel. 5 for the downlink and in Rel. 6 for the uplink, but has no impact on the general architecture here. Another optimization consists in a direct user plane tunnel ("Direct Tunnel") between RNC and GGSN; it was introduced in Rel. 7 and allows, under certain conditions, to bypass the SGSN with the user plane.

The I-WLAN architecture is a separate part of the legacy system, available since 3GPP's release 6 and shown in Figure 2-4 for the non-roaming case. It should be mentioned that the original limitation of this architecture on WLAN as the radio access technology has been relaxed, and it could e.g. be deployed with WIMAX access network in the same manner.

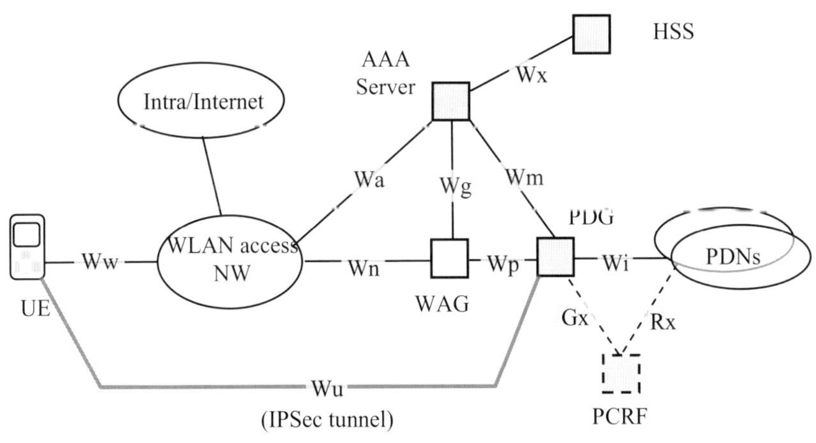

Figure 2-4: legacy I-WLAN architecture (for WLAN access to PDNs via the the 3GPP PS core network; simplified)

The main features here are the presence of a Packet Data GW (PDG), as the point of interconnection towards the PDNs (corresponding to the GGSN), the AAA server (handles authentication/authorization based on IETF protocols), the

IPSec tunnel interface Wu (between UE and PDG), a Wireless Access Gateway (WAG, with some limited, especially routing, functionality) and the local breakout from the local WLAN into an IP domain (Intra- or Internet). The latter shows some similarities with Local IP Access of home cells (see sub-section 3.10.2), whereas the AAA server, an enhanced form of Wu interface and an evolved PDG (ePDG) appear again in the non-3GPP part of the Evolved 3GPP system architecture. PCRF may optionally be present. Only HSS and potentially PCRF are common with the architecture in Figure 2-3, and no mobility between these two accesses is supported (this is in fact a Rel. 8 feature, in parallel to the development of the LTE/SAE system).

3GPP's Rel. 8 adds a completely new air interface and radio access network termed as well as a core network based on packet switching technology. Although these two constitute the largest part of Rel. 8 and are in fact the focus of this book, some improvements in this release are also applied in the legacy part of the 3GPP architecture: e.g. multiple antenna in UE and radio node (MIMO), various HSPA enhancements (for CS voice, higher modulation and MBMS support), improvements in relocation in the RAN (leading to reduced delay and signalling/processing load) and continous packet connectivity.

2.4.2 WIMAX

WIMAX as a broadband wireless technology was developed by IEEE and is available for fixed wireless access since 2001 (802.16.1). The WIMAX Forum Networking Group (WMF NWG) also defined a complete end-to-end mobile network architecture, which became available in its first Release 1.0.0 in March 2007. It is based on IPv4 and the radio interface specification IEEE 802.16e (2005); it supports fixed, nomadic, portable and mobile terminals. This is both a competitor and a subject of integration for 3GPP's evolved system in a cooperative manner, we therefore present a simplified architectural view in Figure 2-5.

The WIMAX terminal is called 'Mobile Station'; on the network side of the air interface the first network node is the Base Station (BS). The Access Service Network Gateway (ASN-GW) concentrates data path and control signaling for a set of BSs. It includes a MIPv4 Foreign Agent or a PMIP client, an AAA proxy and a DHCP proxy.

Layer 2 specific Inter-BS and inter-ASN handover mechanisms support mobility within one ASN and between different ASNs. The core network providing wide area connectivity is constituted by a home and/or visited Connectivity Services Network (CSN). Across the interfaces R3 (between ASN and CSN) and R4 (between visited and home CSN) mobility is supported on the IP layer.

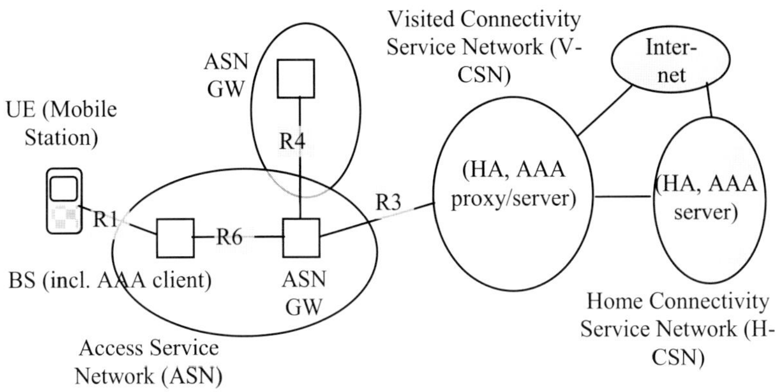

Figure 2-5: WIMAX network architecture (Rel. 1.0.0; simplified)

The WIMAX architecture allows some flexibility regarding distribution of functionality, represented by different "profiles". For our purposes it is sufficient to say that both MIPv4 [9] and a WIMAX specific Proxy MIP protocol (based on IPv4) were supported in rel. 1.0.0. WMF NWG is currently about to add support for IPv6 and MIPv6. One distinct feature of WIMAX networking technology is that it has no legacy, in particular no CS. Thus it is ideally suited for greenfield operators, allows to leverage cost savings from scaling due to all-IP and targets at mobile data-centric usage.

In April 2009 several hundred deployments were claimed in ~130 countries; regarding mobile WIMAX, in June 2009 more than 140 networks were deployed worldwide, with a strong focus on Africa and Central/Latin America. These numbers are expected to increase steadily.

2.4.3 TISPAN / NGN

TISPAN (Telecoms & Internet converged Services & Protocols for Advanced Networks) is a collaboration project focusing on Next Generation Networks (NGN), including fixed network interconnection, evolution and convergence with mobile networks. The simplified TISPAN NGN architecture (Rel. 2 as of 03/2008) is seen in Figure 2-6.

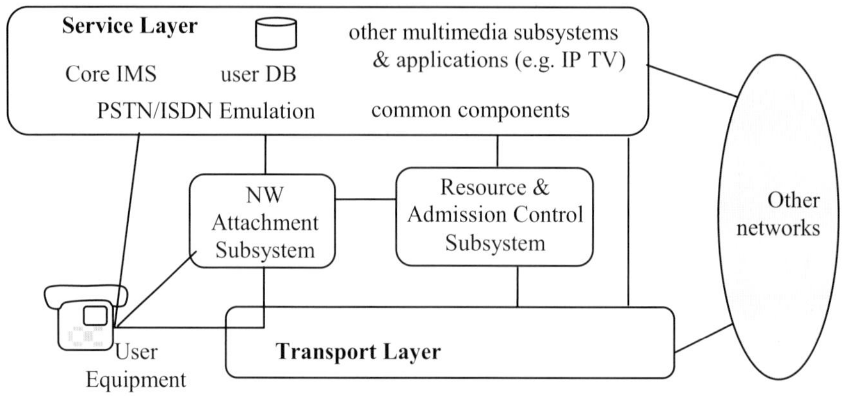

Figure 2-6: TISPAN architecture (Rel. 1; simplified)

The service layer relies on core functionality of IMS as developed originally by 3GPP; 3GPP have enhanced their IMS specifications to cover all TISPAN required functionality. PSTN/ISDN Emulation realizes the well known service functions with IP based control and over IP bearers (in practice telephony over IP, including supplementary) services. Other multimedia services comprise streaming services and content broadcasting, common components include e.g. user profile handling, ENUM resolution, etc.

The Network Attachment Subsystem (NASS) takes care of IP address and terminal configuration, authentication on the IP layer, location management on the IP layer, authorization of network access and access network configuration. The Resource and Admission Control Subsystem (RACS) realizes admission control of user equipment and traffic gating. Authorization checks are performed based on user profiles, operator policies and e.g. resource status. Transport layer functions include layer 2 termination for the access line, access relay functions, gateway functionality for user traffic and signaling towards other networks. TISPAN's functional architecture is specified in ETSI TS 282 001[10] (currently in Rel. 2, with Rel. 3 being under development).

2.4.4 3GPP2 (3rd Generation Partnership Project 2)

3GPP2 is the counterpart of 3GPP for the American, Pacific and partially Asian market. This standardization body has developed also a large set of specifications, describing an own mobile network technology, the current generation being branded as CDMA2000 ©. In some respect they are inspired by 3GPP concepts and solutions, but selectively choose different ones. Regarding LTE/SAE, there has been increasing interest from 3GPP2 operators during the past ~2 years to enable smooth and efficient interworking.

In more detail, the legacy 3GPP2 technology comprises a CS component (called 1xRTT) and a PS component (EV-DO or HRPD). 3GPP2 consider their High Rate Packed Data (HRPD) network as equivalent to 3GPP legacy system; this entitled it for specifically designed, optimized handover procedures. Further evolutions are planned, leading to eHRPD; the extensions within the HRPD network required to support optimized handover with E-UTRAN under the 3GPP EPC umbrella will also be included in eHRPD.

The 3GPP2 network architecture (with MIP mobility) is visible in Figure 2-7 in a simplified manner. The nodes BTS and BSC are comparable to the 3GPP counterparts; PCF is the Packet Control Function. Interfaces A8 and A10 denote the user plane and A9/A11 the control plane interfaces. PDSN is a node comparable to the GGSN in the 3GPP network, as it connects to the IP network cloud (via Pi interface, which is equivalen to the Gi, SGi, Wi and Hi interfaces in 3GPP's architectures). The mobility architecture beyond PDSN is based on Mobile IP (MIP); the AAA infrastructure is is connected by RADIUS protocol interfaces. Simple IPv4 and IPv6 connectivity (i.e. without a MIP Home Agent) and Mobile IP (MIPv4 and MIPv6) is offered.

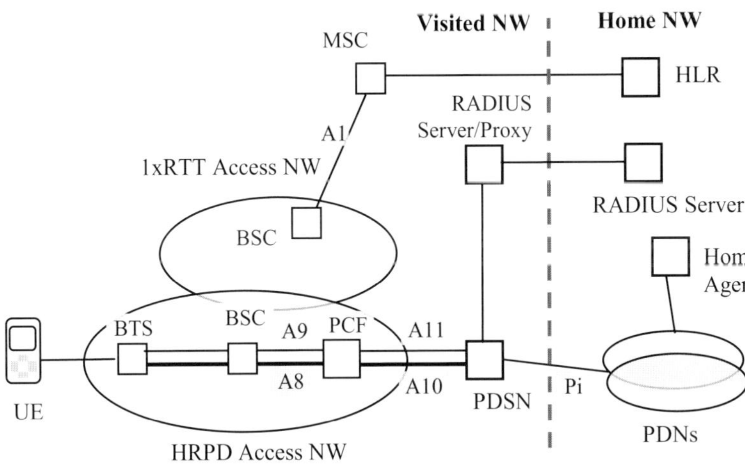

Figure 2-7: 3GPP2's wireless IP network architecture (simplified)

A PDSN upgraded for interworking with the 3GPP EPC is called a HRPD-Serving GW, and thus the node comparable to Serving GW in the evolved 3GPP architecture.

2.5 Battles and compromises: a snapshot in standardization history

It is illustrative, before we present the final, standardized result of 3GPP's system evolution, to look back onto intermediate stages. From the beginning there were two pronounced camps (they surely still exist today, but we will not mention any names here):

- Those companies (operators and vendors) anchored heavily in the legacy mobile network business; naturally, despite their commitment to new technologies and evolution, they pushed for an architecture with the smoothest possible evolution path and minimal impacts on the legacy system. They saw the integration of the new 3GPP radio access preferably in a tight manner (optimized on layer 2), whereas for non-3GPP access technologies they foresaw loose and generic interworking. Subsequently, they kept an eye on preservation or evolution of features present in the legacy 3GPP system; an example for that is SMS in traditional form (i.e. NOT SMS over IP).

- IP friendly companies with new business models, often very active in the non-3GPP domain. Their proposals were for best-available, IP based handover mechanism with non-3GPP access technologies, usage of new protocols and procedures; an example for the latter is Media Independent Handover (IEEE 802.21 [11]), which was finally not taken on board.

A snapshot of how the views diverged intermediately is seen in Figure 2-8; it is the status of two high-level architectural options available on the 3GPP SA2 meeting in May 2005. Not as a surprise, the final architecture of 3GPP Rel. 8 is somewhere in the middle between these two extremes.

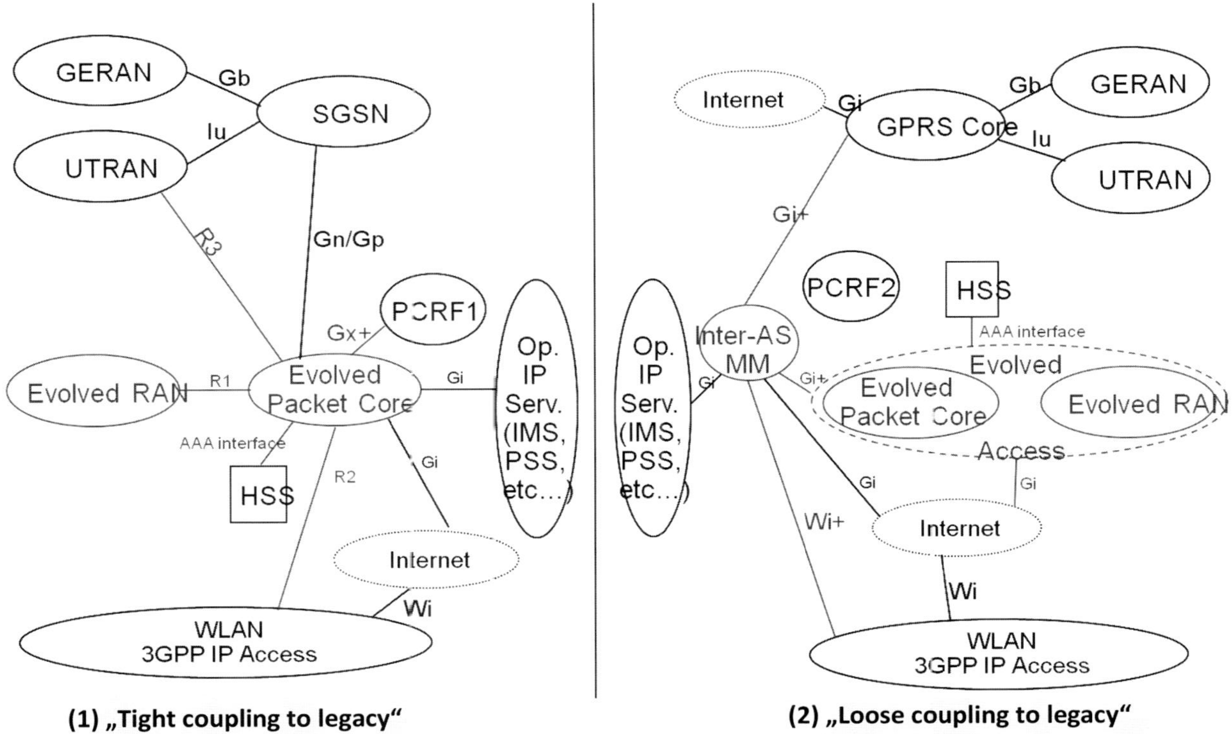

Figure 2-8: two intermediate options of 3GPP's evolved system (May 2005, taken from [5])

This leads us to a concluding side-remark: the whole sense of standardization is to avoid fragmentation of the industry, and so always compromises have to be sought. During design of the Evolved 3GPP system it was often possible, due to the moderating skills of chairpersons in meetings, to proceed based on consensus and to avoid votings on the technical issues; serious consideration of all pros and cons, time allocation for more discussions when a deadlock occurred, and also the graceful behaviour of all 3GPP member companies were and are part of this procedure.

2.6 Commitments for the future

The graph in Figure 2-9 shows the publicly announced number of commitments to LTE/SAE (or in terms of the evolved core network, EPC) by operators at the time of writing (taken from [12]), over the next few years. The numbers are principally available per quarter year, but apparently operators put new infrastructure into operation mostly in the last quarter of a year, or at least announce it in such a way; so there is no value in showing the rest of quarters individually. Although such a statistic is coarse by nature (delays may occur, on the other hand new commitments most likely will be stated), and also does not consider the weight/size of an operator on the market, it gives a good impression of when the heavy rollout of LTE/SAE will likely occur.

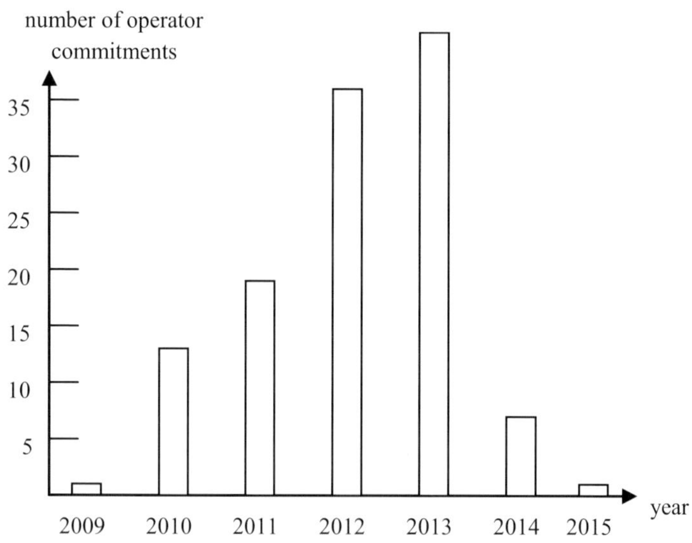

Figure 2-9: commitments of operators for LTE/EPC deployments (as of July 2009)

References

[1] J. Eberspächer, HJ. Vögel: "GSM – Architecture, Protocols and Services", John Wiley & Sons (2009)
[2] A. Kavanagh, J. Beckmeyer: "GPRS Networks", Osborne Publishing (August 2002)
[3] H. Kaaranen, A. Ahtiainen, L. Laitinen, S. Naghian and V. Niemi: "UMTS Networks (Architecture, Mobility and Services)", John Wiley & Sons (2005)
[4] 3GPP TR 25.913: "Feasibility Study of Evolved UTRA and UTRAN"
[5] 3GPP TS 23.882: "3GPP System Architecture Evolution: Report on Technical Options and Conclusions"
[6] 3GPP TS 22.278: "Service requirements for the Evolved Packet System (EPS)"
[7] NGMN White Paper (December 2006), "Next Generation Mobile Networks Beyond HSPA & EVDO"
[8] 3GPP TS 23.002: "Network architecture"
[9] IETF RFC 3344 (August 2002): "IP Mobility Support for IPv4"
[10] ETS I TS 282 001: "Telecommunications and Internet converged Services and Protocols for Advanced Networking (TISPAN); NGN Functional Architecture" (V2.0.0, 2008)
[11] IEEE 802.21, standard to be published (see http://www.ieee802.org/21/)
[12] 3G Americas: www.3GAmericas.com

Chapter 3: Architecture of the Evolved 3GPP System

This chapter presents the architecture of the Evolved 3GPP system, as it corresponds to the 3GPP's stage 2 specifications: that is, the arrangement of functional entities (which may, but need not always correspond to network nodes) and reference points between them. Due to the inherent complexity and variety, this is done in steps: first an overall picture is given; then the evolved radio access network and evolved core network are discussed; finally the architectures for roaming and interworking cases are presented.

3.1 Overall architecture

The overall architecture of the Evolved 3GPP System is shown in Figure 3-1 together with the already existing, 3GPP defined access and core networks (which we call the "legacy 3GPP system"). Also shown are:

- access networks which are not defined by 3GPP but can be used in conjunction with the Evolved 3GPP System (they are called "non-3GPP access networks"); and
- the service domain; it is to be understood as the multitude of IP services, so in general they are represented and realized by Packet Data Networks (PDNs). An IP service may simply offer plain IP connectivity (i.e. enable an Internet connection), provide connection to a corporate network, or feature advanced, IP based control (like telephony or instant messaging via IMS).

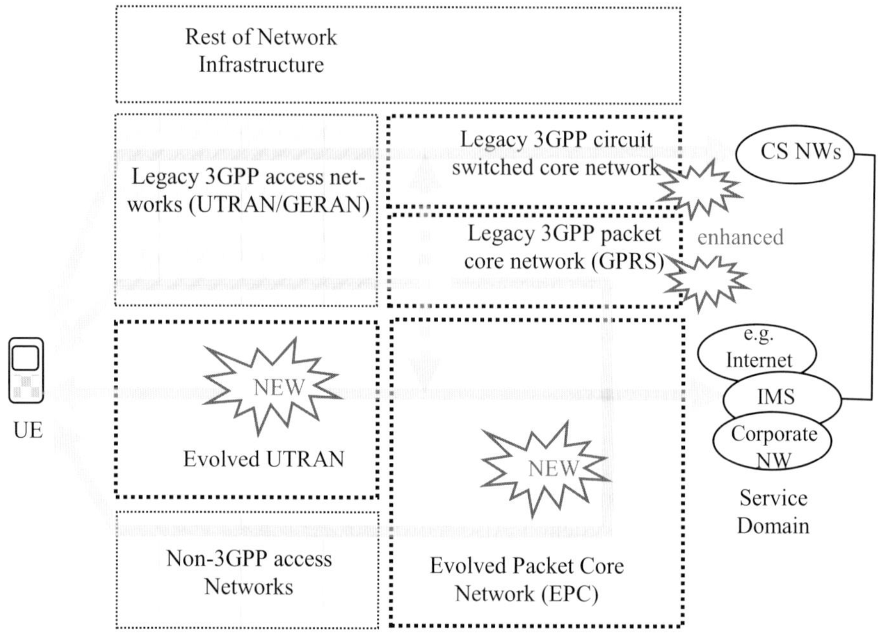

Figure 3-1: Overall architecture of the Evolved 3GPP System

The arrows visualize data paths through these domains; their differentiation into control and user plane (where applicable) is not yet done here, but is the subject of further detailing in subsequent chapters. The evolved network consists of these main parts:

- The new radio access network: it is called "Evolved UTRAN" (E-UTRAN). UTRAN and GERAN are the legacy radio access networks and are connected to the legacy PS domain.
- Evolved Packet Core (EPC): in addition to the basic functions to handle packet routing and forwarding (for the transport of user data), it contains all the necessary control functionality (most importantly for mobility, session handling, security and charging).
- For interworking with the legacy CS domain, the CS core network has to be considered as well and interfaced with the IMS at the backend.

The dotted arrow indicates an optional interconnection between legacy CS core network and the new Evolved Packet Core network, for the purpose of fallback to the CS domain for voice services, if needed (see sub-sections 3.5.3 and 4.14.3).

The user equipment (UE), or terminal, needs of course new capabilities; the most important one is the support of the new radio interface, but also higher level protocols are enhanced, and handling of additional configuration data has to be supported.

Regarding the legacy packet core the principal goal was to minimize impacts there; but it is clear that for optimal interworking with the new system some enhancements are required. Still, there are interworking models based on unchanged (pre-Release 8) legacy PS infrastructure (see sub-section 3.4.3). In the so called "rest of NW infrastructure" some enhancements are obviously necessary (e.g. for the subscription data in HSS, charging system, etc.).

3.2 Evolved Radio Access Network

The architecture of the E-UTRAN is depicted in Figure 3-2. The more general term "Evolved Radio Access Network" (eRAN), may also be used; in the context of signaling protocols also the term "Access Stratum" (AS) may be utilized. The legacy counterpart is also shown here for UTRAN in the insert. The comparison reveals that E-UTRAN is comprised of only one type of nodes, namely Evolved Node B (eNodeB), and that the variety of interconnections has shrunk to a minimum.

eNodeB is the radio base station and transmits/receives via its antenna in an area (cell), limited by physical factors (signal strength, interference conditions, radio wave propagation conditions). It has logical interfaces X2 with neighbouring eNodeBs and with the EPC via S1. Both have a control part (i.e. for signalling) and and a user plane part (for the payload data); for S1 these two are also shown separately in the graph, as they terminate in logical separate nodes within the EPC (the control plane terminates in MME and the user plane in Serving GW, see sub-section 3.3.2). The reference point towards the UE (which comprises the radio link interface and a mobile network related protocol stack) is termed "LTE-Uu", to indicate that it differes from the legacy counterpart Uu.

X2 connectivity between neighbouring eNodeBs can be assumed for most of the E-UTRAN; it is used in most cases of handovers between radio cells, as the UE moves along. Handover preparation is done via signaling across X2 between two involved eNodeBs and user data may be forwarded between them for a short period of time. Details are described in sub-sections 4.3.3 and 5.8.2.

Only in special cases it may happen that X2 is not configured between two neighbouring eNodeBs. In this case handovers are still supported, but handover preparation and data forwarding is then done via the EPC (see sub-section 5.8.3). As a consequence, higher latency and less "seamlessness" must therefore be expected.

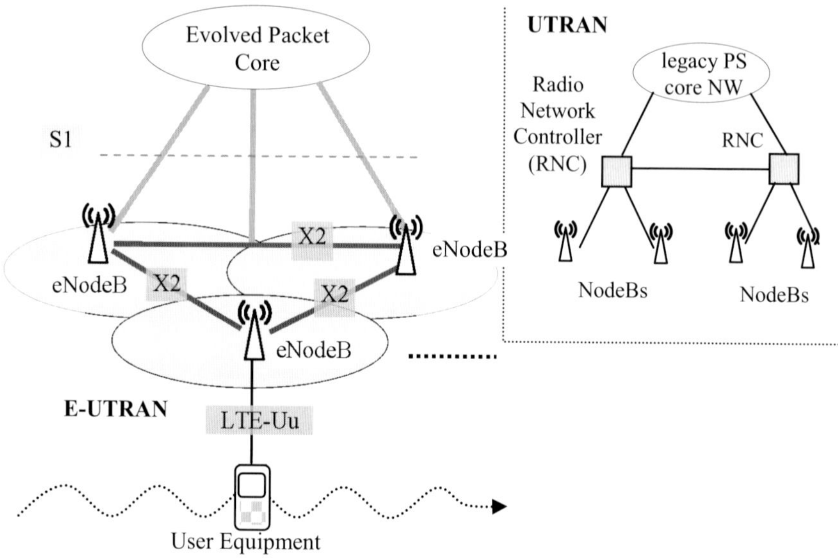

Figure 3-2: Architecture of the E-UTRAN (and comparison with legacy system)

In more detail, the functions performed by the eNodeB are the following:

- Radio Resource Management: Radio Bearer Control, Radio Admission Control, Connection Mobility Control, dynamic allocation of resources (i.e. scheduling) to UEs in both uplink and downlink;
- IP header compression and encryption of the user data stream;
- selection of the control plane node in EPC (MME) at UE's attachment, when no routing to an MME can be determined from the information provided by the UE;
- routing of user plane data packets towards the EPC (in particular, to the Serving GW node);
- transport level packet marking in the uplink, e.g. setting the DiffServ code point, based on the QoS Class Index (QCI) of the associated EPS bearer. (QoS aspects are explained in sub-section 4.11)
- scheduling and transmission of paging messages (upon request from the MME);
- scheduling and transmission of broadcast information (originated from the MME or O&M);
- provision of measurement and measurement reporting configuration for mobility and scheduling.

These functions and related concepts are discussed according to the protocol layer structure in sub-sections 4.1 to 4.3, and some of them are seen in action in the detailed procedures (see sub-sections 5.8.2 and 5.8.3).

☞ The overall architecture of E-UTRAN is defined in 3GPP TS 36.300 [1]; details are contained in the 3GPP TS 36.xxx series.

3.3 Evolved Packet Core Network (EPC)

3.3.1 General remarks on EPC architecture

From the very outset of the architectural work for the Evolved 3GPP system two diverging views regarding realization of mobility with user plane and control plane protocols were present: the first one promoted the smooth evolution of the GPRS Tunneling Protocol (GTP), whereas the other one pushed for new (and what was called "IETF based") protocols. Both had good arguments on their side:

- GTP evolution: this protocol had proved its usefulness and capabilities to operators, and had been extremely successful in large scale operations. It had been designed exactly to the needs of mobile PS networks.
- IETF based protocols: IETF is the de-facto standardization body for the internet. Their mobility protocols evolved from client based Mobile IP towards network based "Proxy Mobile IP" (PMIP). For the internal use within EPC, as it needs to provide network based mobility, only the latter is relevant. PMIP was standardized in parallel to the Evolved 3GPP System. (But note that client based Mobile IP is used in EPS in conjunction with non-3GPP access support.)

Normally protocol selection in 3GPP is seen as a stage 3 issue, i.e. within the detailed specification work following the definition of an architecture; but in this case it was then found that the protocols had profound impacts on the architecture itself, namely the definition of functions and presence of reference points between network nodes. At the end no sharp decision in favour of one and against the other view was possible; the "salomonic" solution taken in December 2006 was to define an architecture encompassing both variants.

In the following descriptions the differences between the two variants is clearly marked both in figures and in text. The architecture of the EPC has also a few instantiations, depending on the roaming situation and on whether interworking with legacy 3GPP accesses or non-3GPP accesses need to be shown. For a better and gradual understanding, we start with the most simple one, and elaborate subsequently.

3.3.2 EPC architecture for 3GPP access in non-roaming

The architectural arrangement for the non-roaming case, without any inter-working with legacy 3GPP accesses or non-3GPP accesses, is shown in Figure 3-3.

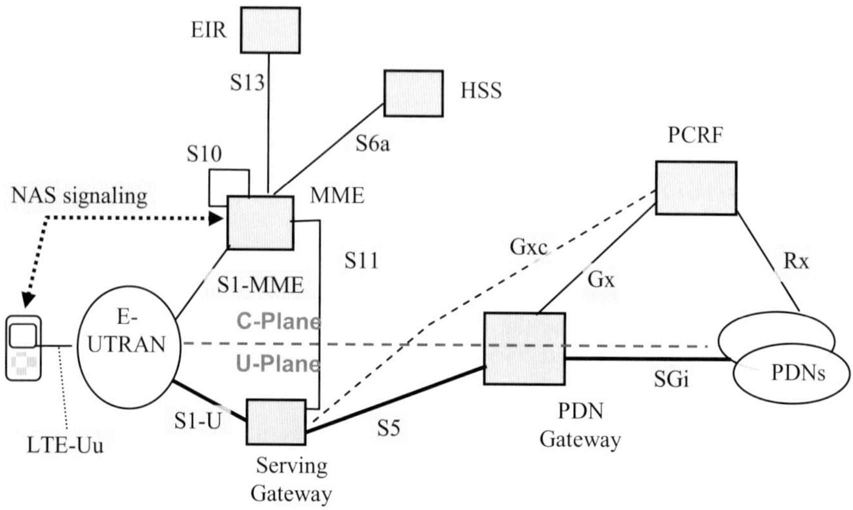

Figure 3-3: Architecture of the EPC for 3GPP access in the non-roaming case (without inter-working with legacy 3GPP and non-3GPP accesses)

Although the logical functions MME, Serving GW and PDN GW are defined separately, they may be implemented in co-locations, i.e. in combined nodes (which then obsoletes S5 interface). In this preferred implementation the user plane, on network level, passes only two nodes within the mobile network infra-structure: eNodeB in E-UTRAN and Serving/PDN-GW (of course this does not count pure transport level nodes like IP routers).

C-(Control) plane and U-(user) plane are visualized in the architecture; MME and Serving GW are genuine representatives of these two different portions of the network, while PDN GW and eNodeb provide functions for both.

The functions provided by the reference points and the protocols employed are:

LTE-Uu: the reference point for the radio interface, between UE and eNodeB, encompasses control plane and user plane. The uppermost layer of the control plane is termed "Radio Resource Control" (RRC). It is is stacked onto "Packet Data Convergence Protocol" (PDCP), Radio Link Control and MAC layers (see sub-section 4.2 for the latter); these three also constitute the user plane.

S1-U: this is the reference point for the user plane traffic between eNodeB and Serving GW. The main activity via this reference point is to transfer IP packets stemming from user's traffic in encapsulated, or tunneled, form. Encapsulation is necessary to realize the virtual IP link between eNodeB and Service GW even during movement of a UE, and thus enable mobility. The employed protocol is based on GTP-U (see sub-section 6.1.3).

S1-MME: this is the reference point for the control plane between eNodeB and MME. All control activities are performed over it, e.g. signaling for attachment, detachment, bearer establishment and modification, security procedures, etc. Note that part of this signaling is transparent to the E-UTRAN, and is exchanged directly between UE and MME; this is part is called "Non-Access Stratum" (NAS) signaling and is detailed in sub-section 6.9. The protocol chosen for S1-MME is S1-AP, on top of SCTP (see sub-sections 6.7 and 6.8).

S5: this reference point includes control and user plane between Serving GW and PDN GW and applies only if both nodes reside in the HPLMN; the corresponding reference point when Serving GW is in VPLMN is called S8. As explained above, two protocol variants are possible here, an evolved GPRS Tunneling Protocol (GTP) and Proxy Mobile IP (PMIP).

S6a: this reference point is for information exchange related to UE's subscription (download and purge). It corresponds to reference points D and Gr in the legacy system, and is based on the DIAMETER protocol (see sub-section 6.5).

SGi: it is the exit point towards PDNs, and it corresponds to the Gi reference point in GPRS and Wi in I-WLAN. Here IETF based protocols for the user plane (i.e. IPv4 and IPv6 for packet forwarding) and control plane protocols like DHCP and RADIUS/DIAMETER for IP address/protocol configuration from external networks are employed.

S10: this is a reference point between two MMEs for the purpose of relocation (i.e. for the case that, due to its movement, a UE is handled by an eNodeB without connectivity to the current MME, or the MME is changed due to other reasons, e.g. load). It is a pure control plane interface and an evolved GTP-C protocol is used.

S11: a reference point for the control plane existing between MME and Serving GW; it employs the evolved GTP-C (GTP-C v2) protocol. The data bearer(s) between eNodeB and Serving GW are controlled via the concatenation of S1-MME and S11.

S13: the reference point connects Equipment Identity Register (EIR) and MME and is used for checking of identity (e.g. based on IMEI, whether they are blacklisted). It uses DIAMETER protocol over SCTP.

Gx: the reference point for QoS policy, filter policy and charging control, between PCRF and PDN GW. It used to deliver filter and charging rules. The protocol used is DIAMETER.

Gxc: this reference point exists in addition to Gx, but is located between PCRF and Serving GW and applies only if PMIP is used on S5 or S8. The reason is that PMIP has no bearer model, thus the so called “binding of bearers” must be done in the Serving GW.

Rx: defined between an Application Function (AF), located in a PDN, and PCRF for exchange of policy and charging information; it uses the DIAMETER protocol.

NAS (Non-Acces Stratum) signaling: this label is not shown in the official architectural graphs, but is mentioned here for convenience; NAS signaling is effectively traveling on top of S1-MME and LTE-Uu, which constitute its lower layer.

The **MME (Mobility Management Entity)** realizes these functions:

- NAS signaling, i.e. handling of signaling directly from and to the UE (which traverses the E-UTRAN transparently); examples of such signaling messages are attach, detach, bearer establishment and modification, authentication requests and responses;
- NAS signalling security: some (but not all, and not in all circumstances) of the messages listed under the previous item are integrity protected and/or encrypted. MME has to do the en/decryption, and checking of the message authentication codes added to the original messages;
- Tracking Area list management: every UE gets allocated a list of Tracking Areas in which it can move in idle without contacting the network;
- selection of PDN GW(s) and Serving GW at initial attach or for re-location purposes; additional PDN GW(s) need to be selected when the UE requests additional PDN connectivity;
- selection of a target MME in case of handovers with MME change;
- subscriber data handling (i.e. signaling via S6a towards the home HSS);
- authentication: challenging the UE and compare with expected result;
- bearer management functions, including dedicated bearer establishment: as already mentioned, MME can be seen as an intermediary node for the control plane between eNodeB and Serving GW. Whenever establishment or modification of bearers is requested, this has to be checked against subscribed and granted limits.
- lawful interception of signalling traffic: this function is needed to fulfill legal requirements. Note that MME cannot intercept user’s data traffic, but only capture events related to users overall activity (e.g. providing the registration state, a rough measure of location, indication of which access is currently used). This data is provided on an interface towards an authority (but not shown in the reference architecture).

The **Serving GW (Gateway)** implements the following functions (for QoS and policy related terms, like QCI, SDF etc., see sub-sections 4.11 and 4.12):

- mobility anchor for inter-3GPP access mobility (i.e. for inter-eNodeB handover and inter-system mobility with the legacy 3GPP system): in this way UE's mobility is shielded from nodes behind the Serving GW;
- support for packet reordering: it sends one or more "end marker" packets to the source eNodeB immediately after switching the path during inter eNodeB handovers;
- downlink buffering of data packets in "idle" mode and initiation of network triggered "service request" procedure (i.e. bringing the UE and the network into "connected" state);
- lawful interception (of user data traffic): this data is provided on an interface towards an authority (but not shown in the reference architecture).
- packet routeing and forwarding;
- for support of QoS in the transport network: packet marking in the uplink and the downlink (DiffServ Code Point in IP packets), based on the QCI of the associated EPS bearer;
- generation of accounting data on user and bearer granularity for inter-operator charging; additionally support of interface to an online charging system.

For the PMIP variant of EPC protocol it implements the bearer binding (see below) and related verification of traffic. It then also has to support DHCP relay and access router functionality (i.e. handling of Router and Neighbor Solicitation, sending of Router and Neighbor advertisements). In roaming, it may also act as a mobility anchor for non-3GPP access (see sub-section 3.6.3).

The **PDN GW** is the node enabling interconnection towards PDNs; the following functions are common for both variants of EPC protocols (see the discussion on protocol variants in the previous sub-section):

- packet inspection and per-user based packet filtering;
- IP address allocation for UEs;
- lawful interception: the collected data is provided on an interface towards an authority (not shown in the reference architecture).
- transport level packet marking in the uplink and downlink (DiffServ Code Point in IP packets), based on the QCI of the associated EPS bearer;
- uplink and downlink service level charging (e.g. based on SDFs defined by the PCRF, or based on deep packet inspection defined by local policy);

- uplink and downlink service level gating control: based on filters (quintuples of source/destination IP addresses and ports, as well as protocol type) the flow of IP packets is allowed or denied (see sub-section 4.12);
- uplink and downlink service level rate enforcement: the amount of IP packets flowing through the node is kept within the applicable limits. Methods to achieve this are rate policing (may lead to dropping of packets) and shaping (e.g. short time buffering), per individual Service Data Flow (SDF);
- uplink and downlink rate enforcement based on the aggregate maximum bit rate for an APN; it applies to all SDFs of the same APN that are associated with Non-Guaranteed bit rate QCIs. The methods are the same as for the above item;
- downlink rate enforcement based on the accumulated maximum bit rates of the aggregate of service data flows with the same guaranteed bit rate QCI;
- DHCPv4 (server and client) and DHCPv6 (client, relay and server) functions for IP address allocation or IP parameter configuration.

For the GTP based architecture, it additionally handles:

- uplink and downlink bearer binding: this is the procedure that associates a service data flow (defined in a PCC and QoS rule based on a service data flow template), to the EPS bearer deemed to transport the service data flow (note: for PMIP variant of the S5/S8 protocol, this is done at the Serving GW);
- uplink bearer binding verification: this is a cross check of the network, if the UE has applied correct uplink bearer binding;
- IP neighbourhood detection functionality: the PDN GW sends Router Advertisements, handles Router Solicitations received from the UEs and handles Neighbour Solicitations. In other words, the PDN GW acts as the access IP router for the UE, terminating the virtual link provided by the GTP-U tunnel.

For MME and Serving GW nodes an area concept may apply, i.e. the operator may assign them geographically; in this case MMEs are grouped into "pools" (which are defined as sets of Tracking Areas). As long as the UE remains within one MME pool area, there is no need to change the MME. MME pool areas may overlap. A similar concept applies for Serving GWs (the term Serving GW service areas is used then; such a service area is again a set of Tracking Areas). These two structurings are independent of each other, most likely they would reflect the average geographical distribution of load in control and user plane, respectively.

3.3.3 EPC architecture for 3GPP access in roaming

In roaming this case the user plane either:

- extends back to the HPLMN (via an interconnecting network), meaning that all user traffic from the UE is routed via a PDN GW in the HPLMN, where the PDNs are connected; or
- for the sake of a more optimal traffic route, it exits from a PDN GW in the VPLMN towards a local PDN.

The first case is called "home routed traffic", the second one "local breakout". (Note that the second term is also used in the discussion off traffic optimization for Home NBs/eNodeB, but with a different meaning – see sub-section 3.10.) Of course, due the very concept of 3GPP roaming, the control plane always involves the HPLMN.

The choice between home routed traffic and local breakout can be made per APN and per user; if more than one APN is used by a UE, different cases might exist in parallel.

3.3.3.1 Roaming with Home-Routed traffic

Roaming with PDN connectivity in the HPLMN utilizes the architecture given in Figure 3-4. It is characterized by the PDN GW being located in the HPLMN, whereas the UE is served by eNodeB, Serving GW and MME in the VPLMN (and of course there is no possibility of co-locations of Serving GW and PDN GW then). This arrangement may be preferred by operators due to the fact that they have direct access to the user traffic, and it corresponds to the majority of cases in the legacy system, where GGSN is located in HPLMN.

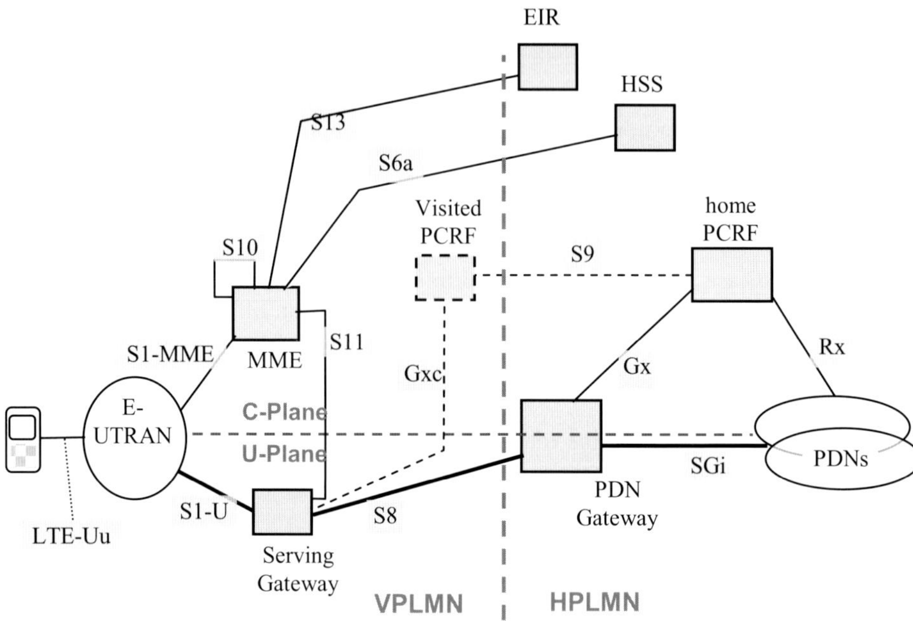

Figure 3-4: Architecture of the EPC in the roaming case with home routed traffic (without interworking with legacy 3GPP and non-3GPP accesses)

S9, Gxc and vPCRF are used only with the PMIP based S8; this is again due the fact to the lack of bearer concept with PMIP, and the need for "bearer binding" in the Serving-GW.

3.3.3.2 Roaming with Local Breakout

The roaming case with PDNs in the visited PLMN follows the architecture presented in Figure 3-5.The difference compared to the roaming case with home routed traffic described above is obviously that the interconnection to PDNs is realized by a PDN GW in the VPLMN. This depends on the support by the VPLMN operator and allows traffic route optimization, if the target PDN is local to the VPLMN. There are two subvariants with respect to PCC: dynamic control of policies and charging is excerted either (1) in the HPLMN, or (2) only in the VPLMN. In the latter case the S9 reference point is needed between HPLMN and VPLMN; this is a new roaming interface.

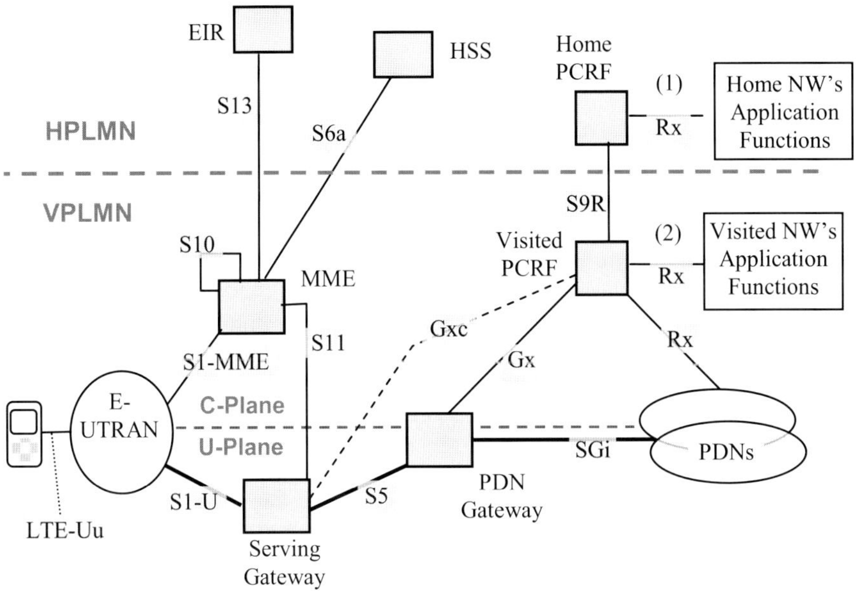

Figure 3-5: Architecture of the EPC for 3GPP access in the roaming case with local breakout (without interworking non-3GPP accesses)

Correspondingly, application functions either (1) in the HPLMN or (2) in the VPLMN may be involved. There is also the possibility that static policies are downloaded by VPLMN from HPLMN via offline, proprietary interfaces, avoiding a tighter coupling and real time signaling.

The advantage here is that services in the VPLMN (in the simplest case just connection to Internet) are accessible in the VPLMN, without a detour via HPLMN. It may result in savings of trunk capacity and delay. On the other hand, it has to be explicitly authorized by the HPLMN (per UE and APN).

3.4 Architecture for interworking between EPC and legacy 3GPP PS system

3.4.1 General

From the very beginning it was clear that the Evolved 3GPP System would have to interwork seamlessly with the widely deployed 3GPP 2G and 3G PS legacy systems – or more precisely, with the GERAN and UTRAN based GPRS. (For aspects of interworking with the CS legacy system, for optimized voice handling, see sub-section 4.14.)

The basic architectural design question is, where to map the GGSN of the 2G/3G in the EPS. 2 variants are available, and both are supported:

1. onto the Serving GW: this is the "normal" case; it means that the Serving GW terminates the user plane (as seen from the legacy GPRS network). The control plane is terminated in the MME, according to the split of user and control plane in EPC. Reference points S4 and S3 are introduced, and they are based on GTP-U and GTP-C, correspondingly. S5/S8 is concatenated towards the PDN GW. The advantage is that the interworking is smooth and optimized; the downside is that for this kind of interworking the SGSN has to be upgraded to Rel. 8 (due to the required support of new functionality on S3 and S4).
2. onto the PDN GW: in this case the unchanged, legacy reference point Gn (in case of roaming it would be Gp) is reused between SGSN and PDN GW, for both control and user plane. The advantage of this usage is that SGSN can be pre-Rel. 8. On the other hand it bears some restriction with respect to IP versions, handover and S5/S8 protocol, as explained below.

For interworking with the legacy system, the MME has to fulfill additional functions (compared to the list in sub-section 2.3.2):

- inter core network node signalling for mobility between 3GPP access networks (terminating S3);
- SGSN selection for handovers to 2G or 3G 3GPP access networks.

In this case the Serving GW additionally has to serve as an anchor for inter-3GPP mobility in the user plane (terminating S4 and relaying the traffic between 2G/3G system and PDN GW).

As a general principle, there is a one-to-one mapping between an EPS bearer and a PDP context. Another problem for interoperation between legacy and evolved system is the mapping of QoS parameters. The detailed rules are somewhat complex and are explained in sub-section 4.11.3.

Another issue is how to minimize signaling in idle mode: a UE in idle mode may choose between E-UTRAN and GERAN/UTRAN also frequently, depending on radio coverage, UE's mobility and preference. For an optimized system behavior, this should not lead to excessive signaling, e.g. for registration. This required a specific solution, and the whole aspect is further discussed in sub-section 4.13.

3.4.2 Interworking with legacy 3GPP PS System based on S3/S4/S5/S8/S12

Based on the complete architecture in Figure 3-3 this is visualized more clearly in Figure 3-6 for the non-roaming case (for roaming the S5 would be replaced by S8). The user plane is shown with thick lines. For this kind of interworking the SGSN has to be upgraded to Rel. 8 (due to the required support of new functionality on S3 and S4). S12 would only be used if the conditions for Direct Tunnel apply.

For this kind of "full" interworking a few enhancements to the legacy SGSN could not be avoided:

- inter EPC node signalling for mobility between 2G/3G and E-UTRAN 3GPP access networks;
- PDN and Serving GW selection: the selection of Serving GW/PDN GW by the SGSN works as it is specified for the MME;
- an MME selection function, for handovers to E-UTRAN 3GPP access network.

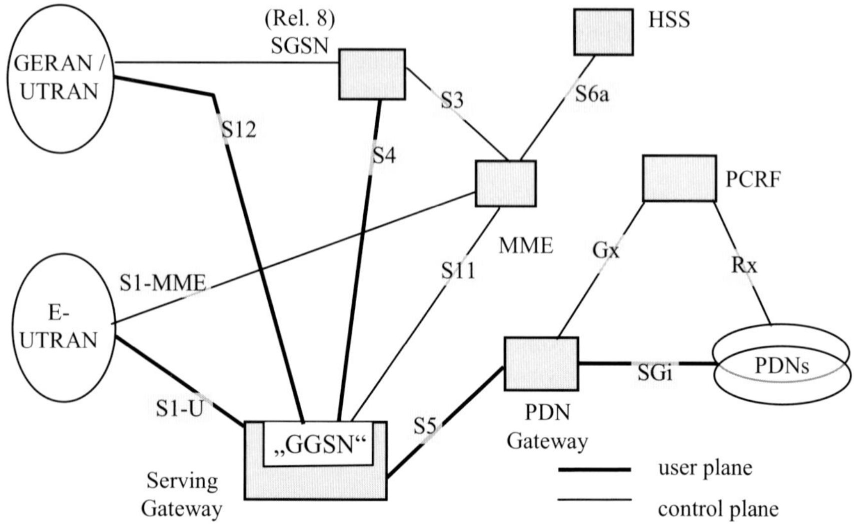

Figure 3-6: S3/S4 based interworking architecture of the Evolved System with the legacy 3GPP system (non-roaming)

These additional reference points were needed:

S3: this reference point realizes control plane information exchange for inter 3GPP access network mobility between the legacy PS system and the EPS.

S4: it provides the user plane tunnelling betwee the legacy PS system and the EPS (with the Serving GW taking the role of GGSN), if Direct Tunnel (S12, see below) is not used. Additionally, it includes control plane for mobility between these two systems.

S12: this reference point is introduced between UTRAN access (in particular the RNC) and the Serving GW, to allow a direct connection of the user plane ("Direct Tunnel"). It is a replica of the arrangement within the legacy system, which was introduced in Rel. 7 and which connects RNC directly with GGSN, bypassing the SGSN under certain conditions for the user plane (only!). GTP-U is used as the protocol here.

3.4.3 Interworking with legacy 3GPP PS System based on Gn/Gp

Figure 3-7 shows (again here only for the non-roaming case) the interworking between legacy and Evolved 3GPP system based on Gn/Gp. In this case the user plane connects via Gn a legacy, pre-Rel. 8 SGSN with the PDN GW, and similarly the control plane via Gn to MME. In effect, PDN GW plays the role of GGSN as "external" point of interconnection for the legacy GPRS network, and MME the role of SGSN in mobility handling (handovers). Gn is then still based on GTPv1 (the legacy protocol variant). The roaming variant of this architecture is readily constructed by substituting Gp for Gn (SGSN, MME, Serving GW are then in VPLMN).

A few restrictions are imposed by this interworking configuration:

- Regarding IP version of bearers: there is no support of dual stack bearers in pre-Rel. 8. Therefore the operator should configure IP addressing so that IPv4 and IPv6 connections use separate EPS bearers. If this is not obeyed, IP addresses would not be maintained when moving from E-UTRAN to a pre-Rel. 8 SGSN.
- If PMIP is used on S5/S8 in EPS, there is no support of handovers between EPS and the legacy system.

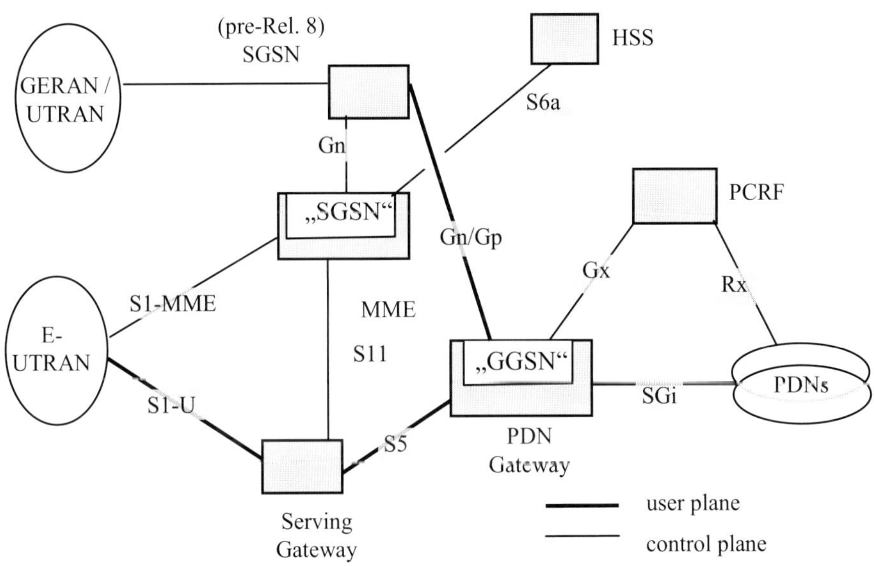

Figure 3-7: Gn/Gp based interworking architecture of the Evolved System with the legacy 3GPP system (non-roaming)

3.5 Architectures for interworking with legacy 3GPP CS System (for voice services)

3.5.1 General

During the design phase of the Evolved 3GPP system, it became apparent that the legacy CS system, with its most important service "voice" communication, could not be ignored by the new system. Operators had simply too much of related investments out in the field, and so high performant interworking was sought. Two solutions were developed:

Single Radio Voice Call Continuity (SRVCC) for transferring ongoing voice calls from LTE (with voice over IMS) to the legacy system, and

CS fallback (enabling a temporary move to the legacy CS before a CS incoming or outgoing activity is performed).

A third one was brought into discussion, but not standardized; it proposes to tunnel CS signaling over the EPC, see [2].

The legacy CS system is to be understood as 3GPP's GSM/UMTS or 3GPP2's 1xRTT.

3.5.2 Architecture for SRVCC (Single Radio Voice Call Continuity)

Figure 3-8 shows the interworking architecture for SRVCC (for simplicity Serving GW and PDN GW are not shown separately). In this solution chosen by 3GPP for SRVCC with GERAN/UTRAN, a specially enhanced MSC (actually its control part, i.e. MSC server) is connected via a new control plane interface (Sv reference point) to MME. Note that the MSC serving the UE may be different from the one supporting the Sv interface (in this case the serving MSC relays towards the SRVCC capably MSC). In the IMS, an application server (AS) for SRVCC is required. Sv is based on GTPv2 and enables preparation of the resources in the target system (access and core network, plus interconnection between CS and IMS domain), while still being connected to the source access.

Similarly, SRVCC with CDMA 2000 © 1xRTT requires the 1xRTT Interworking Server (IWS), which supports S102 interface and relays signaling to/from the 1xRTT MSC serving the UE, with the same purpose as described above. S102 is a tunneling interface and conveys 1xRTT signaling messages; between MME and UE these are encapsulated.

☞ The architectural concept for SRVCC is defined in 3GPP TS 23.216 [3].

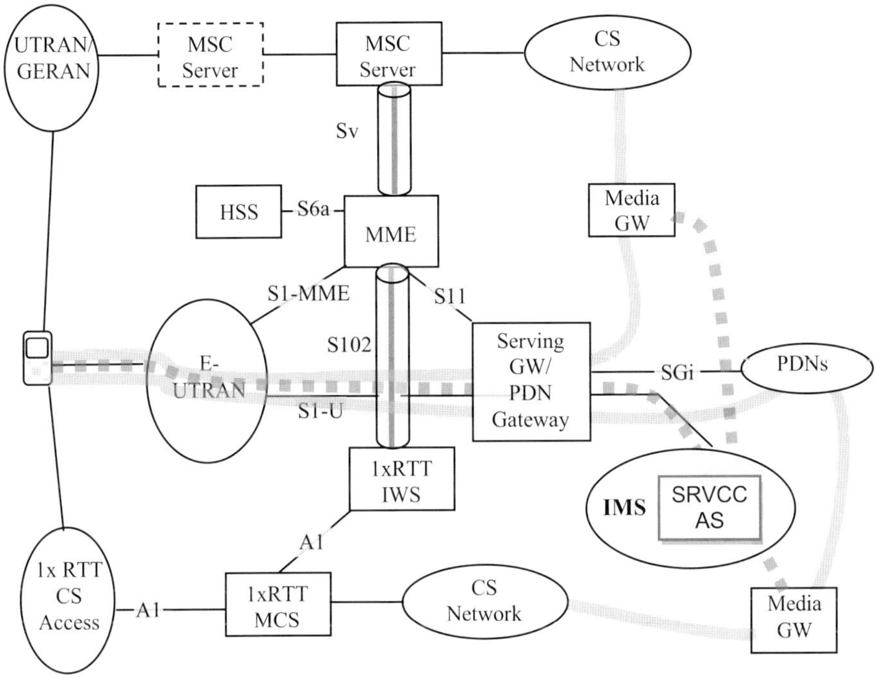

Figure 3-8: architecture for SRVCC between E-UTRAN and GERAN/UTRAN (upper part) and CDMA 2000 © 1xRTT (lower part) (simplified)

More details are explained in sub-section 4.14.2 for the underlying concept and sub-section 5.13 for signaling message flows.

3.5.3 Architecture for CS Fallback

The network architecture for CS fallback from E-UTRAN to GERAN/UTRAN is depicted in Figure 3-9; for simplicity Serving GW and PDN GW are not separated (S5/S8 is thus not exposed), and the VLR is shown integrated with the MSC Server. A new interface SGs is introduced between the MSC Server/VLR and MME, which allows combined and coordinated procedures. The concept consists of:

(A) relaying signaling for terminating CS request (incoming calls, network triggered Supplementary Service handling or legacy SMS) from the MSC Server to the MME over SGs, and vice versa; and

(B) running combined procedures (attach, tracking area update, detach) between the PS domain and the CS domain.

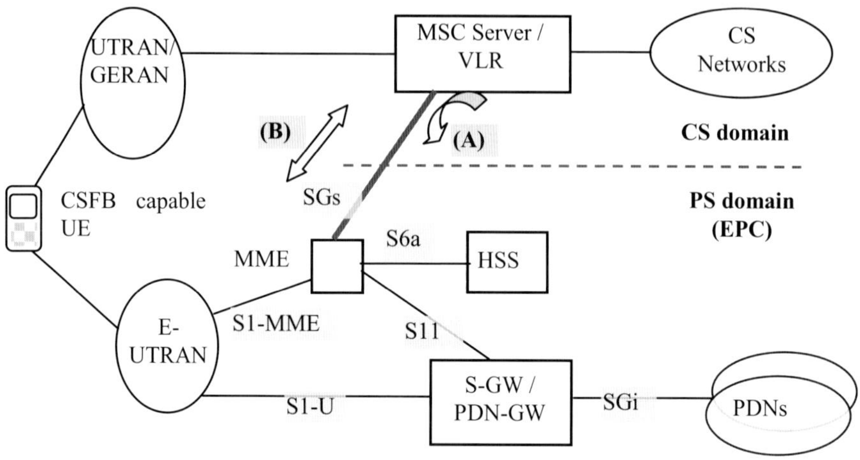

Figure 3-9: Architecture for CS Fallback to GERAN/UTRAN

A corresponding architecture for CS fallback to CDMA 2000 © 1xRTT was defined (Figure 3-10) and is fully in line with the one used for SRVCC.

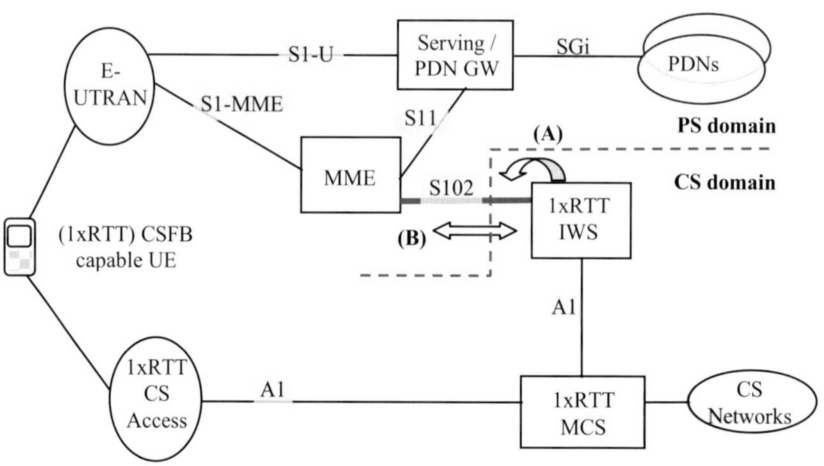

Figure 3-10: Architecture for CS Fallback to CDMA 2000 © 1xRTT

In this case the S102 interface is reused for the CS related control messages. More details are found in chapter 4 (on the related concept) and chapter 5 (on signaling message flows).

3.6 Architecture for Interworking with non-3GPP Accesses

3.6.1 General

Interworking with systems/access networks different from 3GPP's (called non-3GPP systems/accesses) was an important target for SAE; this should happen under the EPC as the "umbrella" (see sub-section 2.2.5). Such interworking may be realized on different levels (and in fact this was done on layer 4 with VCC/SRVCC, see sub-section 4.14.2). But for the generic type of interworking it seemed necessary to rely on generic mechanisms, and so the IP level seemed most appropriate.

In general, "complete" mobile or fixed network systems have a similar architecture as outlined above for the Evolved 3GPP system: they normally consist of an access network and a core network. In the intended interworking architecture of the Evolved 3GPP system, access systems of other technologies connect to the EPC.

It was also decided to allow 2 different types of interworking, depending on the trust property of access systems. For trusted non-3GPP access networks it is assumed that secure communication between them and the EPC is implemented, and also strong enough means of data protection are guaranteed. How this is achieved is not specified, and the categorization of a non-3GPP access network being "trusted" remains fully an operator's decision. In practice, typical examples of trusted non-3GPP access networks are WIMAX, fixed network access (e.g. according to TISPAN) architecture and CDMA 2000 © HRPD. For untrusted non-3GPP access networks no particular assumptions hold (although they may offer some level of security); a typical example is a WLAN.

3.6.2 Non-roaming

The architecture for interworking with both types of non-3GPP accesses in the non-roaming case is seen in Figure 3-11. ePDG is an evolved Packet Data Gateway (enhanced from PDG, as it was used in I-WLAN since Rel. 6) for network based mobility; it realizes the Mobile Access Gateway (MAG) function of PMIPv6 (see sub-section 4.15.2). The PDN GW has now become the anchor for IP based mobility between EPC and non-3GPP accesses.

In more detail, the functions of ePDG include:

- allocation of remote IP address (it is local to the ePDG and is used as CoA when S2c is used);
- transport of IP address information if S2b is used;

- IPSec tunnel handling (optionally including tunnel update due to UE mobility between untrusted non-3GPP accesses);
- PMIPv6 signaling (including GRE key handling);
- tunnel authentication and authorization (based on IKEv2 and AAA signaling);
- en/decapsulation of IP packets and routing through tunnels (to/from the IPsec tunnel and to/from theh PMIP tunnel);
- transport level packet marking in uplink direction;
- enforcement of (static) QoS (if a subscriber specific QoS profile is received from the AAA infrastructure);
- lawful interception.

The PMIPV6 related functions are not used if S2b is not used and S2c is used instead for realizing mobility.

For PDN GW we have, in addition to functionality present for 3GPP accesses, the signaling and tunneling capabilities of:

- Local Mobility Anchor (LMA) function according to PMIPv6 (see sub-section 4.15.2 and [4]),
- Home Agent function according to DSMIPv6 (see sub-section 4.15.3 and [5]),
- Home Agent function to MIPv4, if S2a is used with MIPv4 Foreign Agent CoA mode (see sub-section 4.15.4 and [6]).

An additional set of reference points is therefore introduced:

SWu: it uses IKEv2 for control signaling and IPSec for the user plane; it ensures protected traffic through the untrusted non-3GPP access network.

STa/SWa: reference points between the AAA server and an trusted and untrusted non-3GPP access, respectively, for authentication and authorization of the UE. They use DIAMETER and EAP-AKA on top of it.

S2a/S2b: reference points for user plane and network based mobility signaling and are using PMIPv6; additionally, S2a allows also MIPv4 in FA mode.

S6b: is used for mobility related authorization at the DSMIPv6 Home Agent and the PMIPv6 LMA.

S2c: is the reference point directly between a UE and the Home Agent function (within PDN GW), which is intended for client based mobility by virtue of the DSMIPv6 protocol. The DSMIPv6 mobility tunnel would go through any non-3GPP access network (but it is shown independent from these, for clarity); in order to avoid tunneling overhead, it was defined that within 3GPP access(es) the UE is at home in the DSMIPv6 sense (in this case also no mobility binding is active, see sub-section 4.15.3).

Gxa: this is a reference point for policy interactions between the trusted non-3GPP access network and the PCRF entity in the EPC. Similarly, Gxb is intended for delivering policies from PCRF to ePDG, but it is not specified in Rel. 8; it can be understood as a hook for later enhancements.

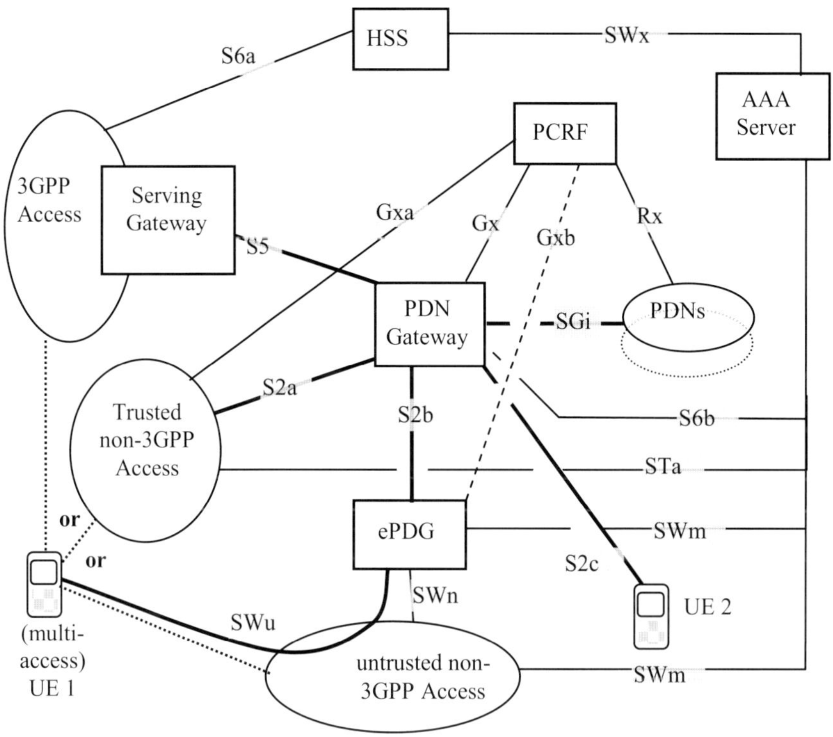

Figure 3-11: Architecture for interworking with non-3GPP access (non-roaming)

Note that the UE cannot permanently be connected via several (3GPP or non-3GPP) accesses to the same PDN; see the limitations of 3GPP's Rel. 8 in subsection 3.13.

3.6.3 Roaming

In roaming with home routed traffic, the architecture is expanded by placing the PDN GW into the VPLMN, introducing vPCRF (PCRF in the VPLMN) and connecting it to the PCRF in the HPLMN, and placing a AAA proxy in the VPLMN and connecting it with the AAA server in the HPLMN.

A variant is constituted by connecting the ePDG not with PDN GW in HPLMN (Figure 3-12), but with a Serving GW in the VPLMN; this "chaining" scenario is only possible if the VPLMN has a business relationship with the (untrusted) non-3GPP network. It allows the Serving GW in VPLMN to include local non-3GPP anchor functionality (however, in rel. 8 there is no standardized way to guarantee the preservation of the Serving GW address in all mobility scenarios). The UE is then preferably handled locally for mobility across non-3GPP accesses in the VPLMN (at least for the user plane).

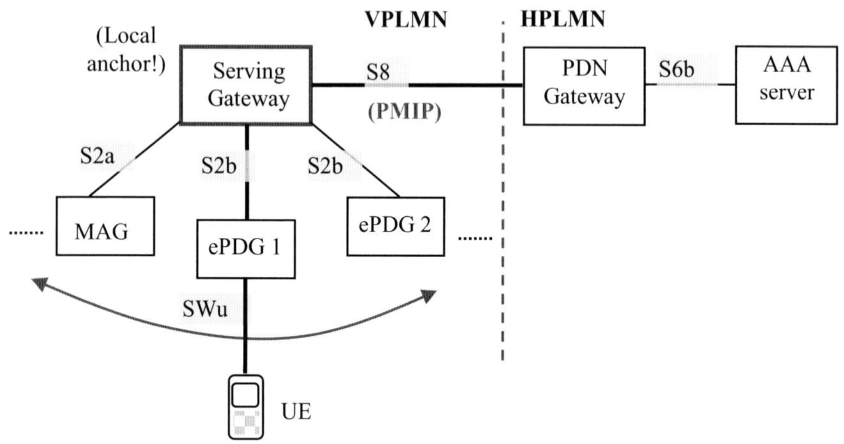

Figure 3-12: principle scheme of chained S2a/S2b in roaming with home-routed traffic

3.7 Architecture for optimized handover with CDMA 2000 © HRPD

CDMA2000 © HRPD (High rate Packet Data) is seen as a PS domain belonging to the legacy mobile network technologies; yet, with respect to the user plane and mobility signaling interfaces it is treated as a trusted non-3GPP access, utilizing PMIPv6 on S2a.The architecture in Figure 3-13 (for the non-roaming case) enables optimized handovers by two specialized interfaces: S101 for handover control –(preregistration and signaling for handover preparation) and S103 for packet forwarding in downlink during handover from E-UTRAN to HRPD, in order to minimize packet loss.

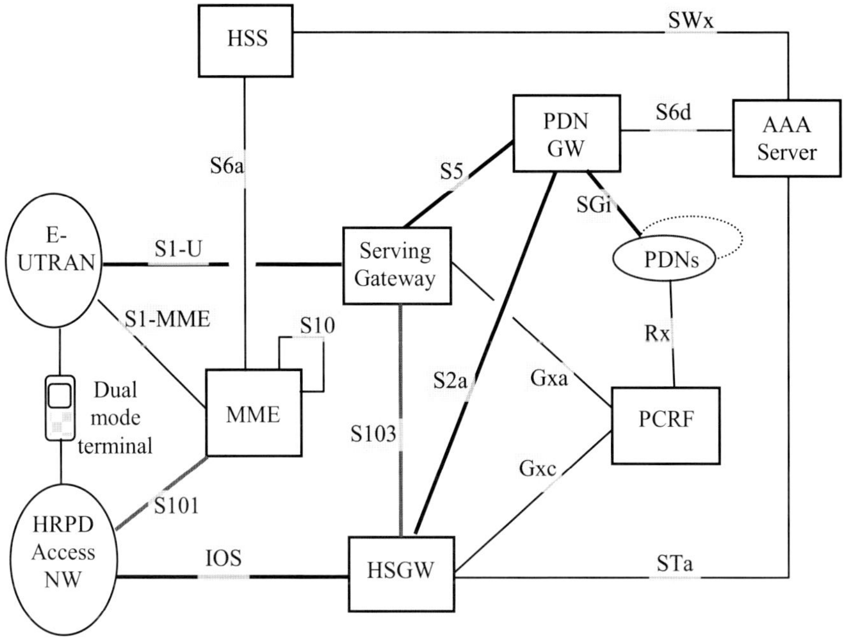

Figure 3-13: architecture for optimized handover with HRPD

S101 is a tunneling interface; signaling messages pertaining to the one access technology are encapsulated and transported through it, while the dual mode UE is attached to the other access technology. They are not interpreted by the MME or the HRPD access network. Reference point Gxc is used only with PMIP based S5.

3.8 Architecture for I-WLAN mobility

The intention behind introduction of this specifically limited architecture was to have an early predecessor solution before LTE/SAE would be in place, enabling mobility between the legacy systems GPRS and I-WLAN. It is compatible with the fully fledged SAE/EPC architecture and builds on DSMIPv6.

In Figure 3-14 the Home Agent is visible as the central mobility enabling component. It acts as the anchor for mobility between the legacy PS accesses I-WLAN and GPRS, together with the newly introduced reference points H1, H2 and H3 (already existing reference points of the legacy system are not shown). See subsection 4.16 for a more detailed explanation of the concept.

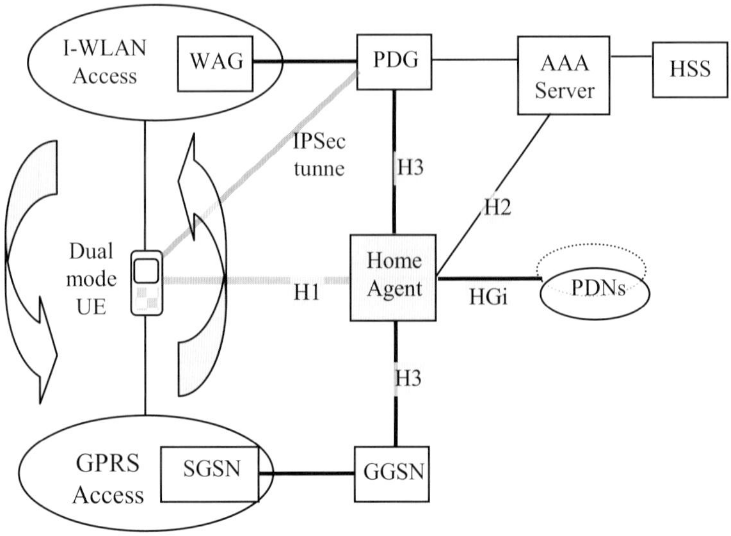

Figure 3-14: Architecture for I-WLAN mobility

HGi is a standard IP interface; it corresponds to Gi in the legacy GPRS, Wi in legacy I-WLAN and SGi in the EPC architecture. H1 is the reference point representing the DSMIPv6 signaling relationship (including security) and user plane tunnel directly between the mobile terminal (UE) and the Home Agent. H2 is a reference point necessary for the authorization of the mobility and charging. H3 is a reference point representing simply IP connectivity (i.e. only packet forwarding).

Note that both the IPSec tunnel (between UE and PDG) and the DSMIPv6 tunnel (between UE and HA) encapsulate the user traffic (i.e. DSMIPv6 tunnel is inside the IPSec tunnel; the same applies for the PDP contexts when the UE is in GPRS access).

The Home Agent realizes the mobility between the two legacy accesses; it may reside in the HPLMN or in the VPLMN.

3.9 Architectural enhancement for access network discovery and selection

As shown in the preceding section, interworking with non-3GPP accesses is an important feature of the Evolved 3GPP system; the number of them available in parallel, and their nature were not to be restricted from the very outset. Soon it was detected that this presents a new challenge: how to find always the "best" or most suitable access network?

A comparison with operation in (and thus selection between) only 3GPP accesses reveals that the situation is different if non-3GPP accesses are considered (see Table 3-1).

	3GPP accesses	Non-3GPP accesses
1	deployment planned/coordinated by 3GPP operator	(normally) no planning/coordination for/by 3GPP operator
2	full configuration data available across overlapping access networks (neighbor cell lists)	no configuration data available across access network boundaries
3	access topology relatively stable	access topology may change frequently and drastically (e.g. hotspot operators may join or leave co-operations)
4	dual radio not possible (due to band limitations, e.g. close bands)	dual radio operation mostly possible (due to diverse bands)
5	handover procedures optimized and realized on L2	handover procedures (mostly) not optimized and realized on L3
6	UE guided by specialized and standardized policy framework (lists of preferred/forbidden accesses & networks)	Before 3GPP's rel. 8: standardized policy framework missing

Table 3-1: Comparison between 3GPP and non-3GPP access selection

At that point two options were still open:

1. Expand 3GPP's static configuration framework for more radio access technologies: this option would have required to replicate all sorts of lists in USIM, HSS and in transport protocols between intermediate nodes. It would not have been generic and flexible enough (e.g. list sizes in data stores and in message buffers would have been critical).
2. Let the UE act rather autonomously (scan and detect access networks, try access if seen advantageous): this strategy is still valid, if nothing else is done.

Based on this analysis it was concluded that the architecture should be enhanced with an optional functional entity within EPS, capable of storing flexibly diverse data for access network discovery and selection. This was called "Access Network Discovery and Selection Function" (ANDSF), and the only interface defined in Rel. 8 is towards the UE represented by reference point S14 (see Figure 3-15).

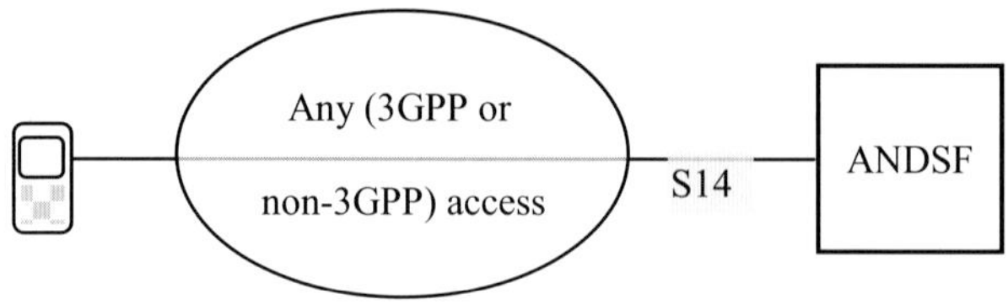

Figure 3-15: Architecture for ANDSF

ANDSF is essentially a database containing policy data for non-3GPP access related inter system mobility, and data for discovery of non-3GPP access networks. S14 utilizes the OMA DM protocol [7], with a 3GPP specific Managed Object definition; the alternative proposal at the time for protocol selection was Media Independent Handover/IEEE 802.21 protocol [8], but was at that time deemed not to be mature enough.

How ANDSF is populated with data is up to the operator; although there was a strong desire to introduce interfaces to other nodes in the EPC (e.g. HSS). This was not accepted for Rel. 8. Also, in Rel. 8 nothing else is specified for the roaming case than the rule that ANDSF must always reside in the HPLMN. This means, S14 is not specified as a roaming interface (between two PLMNs); in particular, there is not yet the notion of "home" (H-) and "visited" (V-) ANDSF. The topic is followed on in Rel. 9, and it requires to elaborate on two sensitive questions:

1. How do the two ANDSFs interact: can e.g. V-ANDSF filter out information sent from H-ANDSF?
2. How can it be avoided that information sent from H-ANDSF via V-ANDSF to the UE is not open to the roaming partner?

Rel. 9 will tackle these issues and bring along a solution based on some simplifying assumptions.

Further details are found in sub-sections 4.15.10 (on the concept of ANDSF), 5.12 (on example message flows) and 6.10 (on the protocol used).

3.10 Architecture for Home Cell Deployment

3.10.1 3GPP Release 8

During development of the Evolved 3GPP system (Rel. 8), operators brought up the strong wish to build in enhancements for support of home cell deployments. This was clearly a countermeasure against the huge amount of deployed WLANs (private and public). The idea is to provide (necessarily cheap) radio equipment for operation in operator's licensed bands to mainly private, but potentially also enterprise customers, install them in homes and enterprise premises and integrate them via the anyway existing fixed (prefereably broadband) access lines into the operator's existing mobile network infrastructure. The expectations are:

- huge mass market,
- competitive service offering,
- cheap extension of e.g. in-house coverage,
- minimal burden on operator's existing infrastructure (no planning, re-use of backhaul connections),
- integration with in-house networks, and
- offload of traffic from operator's radio network (in case of local breakout also from the core network).

The feature of Closed Subscriber Groups (CSGs) is discussed in sub-section 4.19. Here we present the basic architectural configurations.

The term "femto cell" was soon coined for such kind of deployments; the standard, non-home cells are then called macro-cells (which might cause some confusion, because the categories of cell granularity macro/micro/pico may also be used within a pure operator deployment). For the LTE radio technology the radio node is called "home evolved Node B" (Home eNodeB); for legacy radio nodes (but only UMTS radio technology) it is a "home NodeB".

Different architectural variants exist, according to whether the UE is pre-Rel. 8 or Rel. 8 capable, and whether it should be enhanced with specific home cell features, like CSG, or not. Another differentiation of architectures comes into play, depending on whether a gateway further back in the network is used as a concentrator or not. Figure 3-16 gives the Rel. 8 architecture for home cell deployment based on Home NodeB, i.e. based on UTRAN radio access technology.

A new reference point Iuh is introduced, for the signaling performed by Home NodeB towards the Home NodeB GW on behalf of the UE; it is based on the Iu protocol. From there the legacy signaling protocols are used towards the core network nodes (using existing Iu-CS and Iu-PS protocols, see sub-section 2.4.1).

The C1 reference point foresees an interface between a server in the operator's network and the UE; it is used optionally by the operator to configure UEs with their allowed (white) list of CSGs (additionally users can try to use CSGs, which are displayed on their terminal, and if successful i.e. allowed by the network, these will be stored in their allowed CSG list). The C1 interface is based on the OMA DM protocol or on the 3GPP defined over-the-air (OTA) provisioning (which is actually utilizing special SMS). The CSG membership is also known to HSS and will be retrieved by the core network node (SGSN, MSC) whenever an HSS interrogation takes place. No standardized interface between CSG Server and HSS has been defined, but it would be more or less required for a full, larger scale deployment.

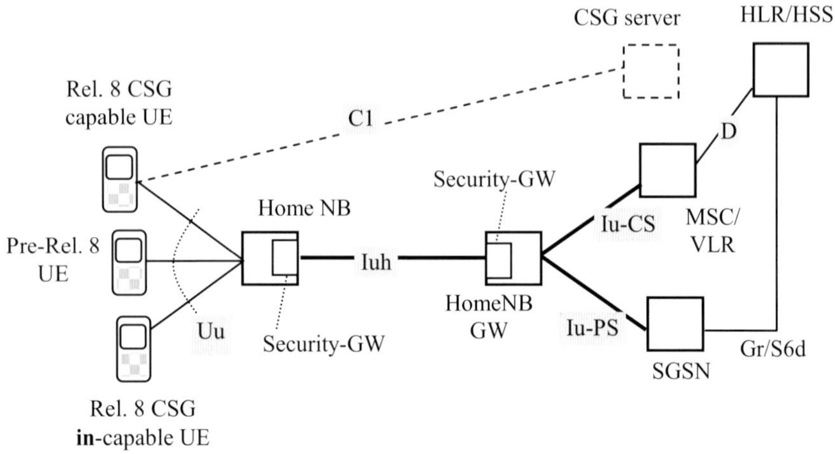

Figure 3-16: Architecture for Rel. 8 home cell deployment based on Home NodeB

The Home NodeB is enhanced from a normal NodeB by mainly these aspects:

1. optionally: sending information on the radio interface regarding to which Closed Subscriber Group it belongs (only one or none);
2. supporting the interface towards the Home NodeB GW by re-using the Iu protocol;
3. acting on behalf of pre-Rel. 8 UEs (registering them with the Home NodeB GW and keeping their context).

Pre-Rel. 8 UEs have naturally no knowledge of neither the Home NodeB nor CSGs. Because of their legacy signaling behavior, they are handled by Home NodeB and Home NodeB GW in such a way that they remain unaware of the fact that their radio link is towards a Home NodeB.

Rel. 8 UEs are aware that the radio station is a Home NodeB and not a normal NodeB; however, they may either support CSGs or not. In the latter case they will not access CSG cells, and no CSG membership can be administrated for them over C1.

The security GWs may be integrated into the HomeNB GW and HNB (as shown), or physically separated. The HomeNB GW acts as a concentrator for HNB connections from customer premises towards the operator network.

The Rel. 8 architecture for home cell deployment based on LTE/HeNodeB/EPC is shown in Figure 3-17.

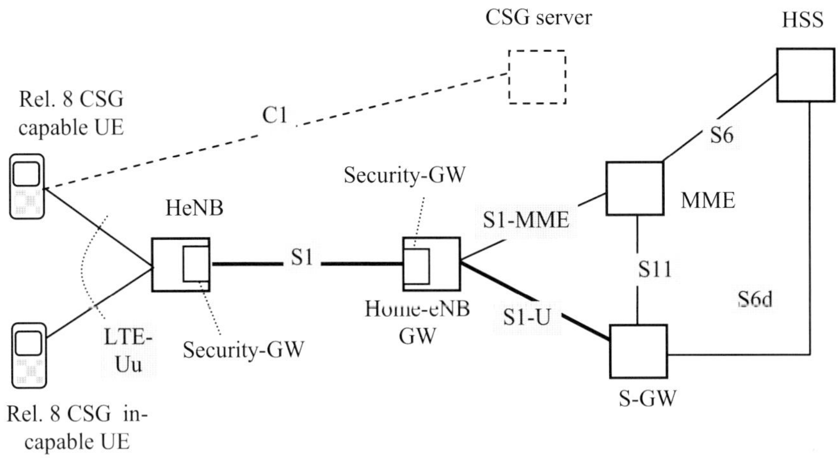

Figure 3-17: Rel. 8 architecture for home cell deployment based on HeNodeB (with Home eNodeB GW)

Two other architectures are defined, but not shown graphically here:

1. without Home-eNodeB GW: in this case Home eNodeB connects directly via S1-MME to MME and via S1-U to the Serving GW;
2. with Home-eNodeB GW only for the control plane: then the Home eNodeB connects to the Serving GW directly via S1-U, but to Home-eNodeB GW and further on to MME via two concatenated S1-MME interfaces.

The mobility between home cells and macro cells is restricted in Rel. 8, due to the complexity in the radio access network and the short timeline given for standardization; inbound mobility (from macro cells towards a home cell) is not supported.

3.10.2 3GPP Release 9 and 10 (Preview)

Several extensions of the home cell concept are foreseen for Rel. 9:

1. Further elaboration of the time based membership in CSGs: in order to improve user experience and CSG control, the UE receives the information on a validity timer (in Rel. 8 it is only known in the network, so the UE could try unnecessarily to make access).

2. Two Allowed CSG lists: one is to be populated by the user (when doing manual mode CSG selection); the other one is for administration by the network operator, and for storing updates when automatic CSG selection happens.

3. Open and hybrid mode: in open mode H(e)NB cell does not broadcast a CSG identity, thus CSG membership is not required for access. In hybrid mode the CSG identity is still broadcast, but UEs without and with matching CSG membership may access; the latter then receive preferential treatment with regards to the available resources.

4. Local IP Access (LIPA): this mode of connectivity is a shortcut for a user's traffic to/from her own home IP network while in her homecell. (The scenario can of course be broadened to campus, with a local IP network.) A very similar access mode was defined for a UE accessing the local/home IP network when away from the home cell, but it was termed "Remote Managed Access". The requirements for LIPA in Release 8 did not yet mandate service continuity, but it can be anticipated that LIPA and Remote Managed Access can be unified when imposing service continuity over home cells, macro cells and non-3GPP access.

Regarding open/hybrid mode, there are no impacts seen on the overall architecture, but some enhancements in signaling (for appropriate access control) are necessary.

LIPA is subject of a study work in Rel. 9 but will come to full bloom with Rel. 10; it anticipates the new target traffic scenarios (1) to (4) in Figure 3-18, in comparison to the standard case (5); for generality two home cells are shown, but only one acts as the anchor for LIPA traffic. For LIPA, as of time of writing, 3GPP had not arrived at a conclusion for the architecture.

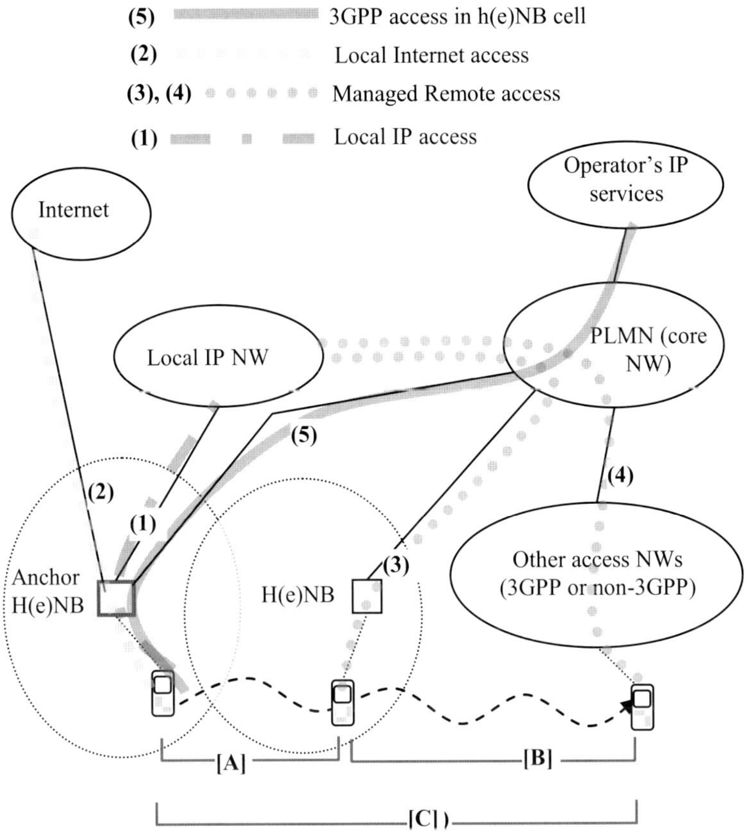

Figure 3-18: target connectivity scenarios for LIPA (2 home cells, but only one as anchor)

Mobility scenarios when the UE moves away from the home cell ([A], [B] and [C] in Figure 3-18) with LIPA are currently not strictly required, but likely to come. The differentiation of LIPA traffic from other, normal traffic is envisaged by specific APNs.

At the time of writing the LIPA concept is also strongly required for macro cells (naturally only for local access to internet), and the architectural design is under heavy debate. Operators see the chance to flatten even more their network and offload "dumb", bit-pipe traffic (which requires no value-added handling) as soon as possible by routing it to the nearest ISP. In this case there are operator requirements to enable the local routing of traffic selectively also within the same APN.

3.11 Architecture for Warning System (ETWS)

The warning system architecture in Rel. 8 was actually a solution for the special requirement of an Earthquake and Tsunami warning system (ETWS) for Japan. For the E-UTRAN part the architecture looks most simple and utilizes the existing interfaces between UE and MME in the control plane (see Figure 3-19). Additionally the MME in EPC is connected to the Cell Broadcast Center (CBC) via the reference point SBc. Warning messages originate in a Warning Center, which acts as Cell Broadcasting Entity (CBE). For the legacy radio accesses the Cell Broadcast Service (CBS) architecture was reused.

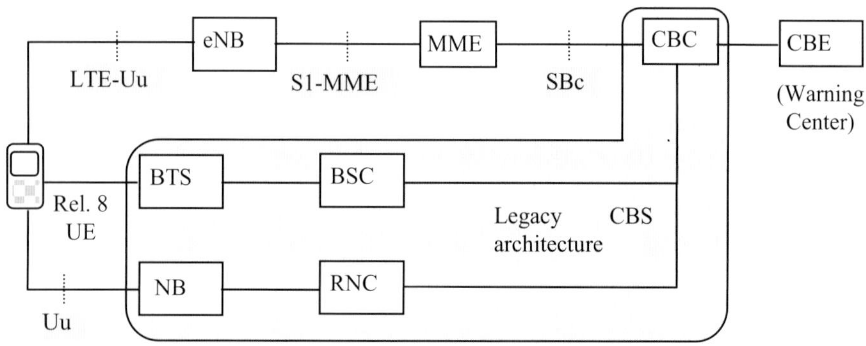

Figure 3-19: architecture for warning system (ETWS)

The functionality of a Cell Broadcast Center has been defined since long for this service, a variant of SMS for point-to-multipoint delivery. The interfaces between CBC and CBE, and between CBC and BSC/RNC are not standardized; only the logical structure of information transferred is specified.

The new SBc protocol, between CBC and MME, is fully standardized and based on SCTP (see sub-section 6.6). S1-MME is also specifically enhanced for the forwarding of the warning message. The message flow for warning message delivery can be found in sub-section 5.15.

3.12 Examples of deployment/operation

At this point, before introducing the concepts of the Evolved 3GPP system in more detail, it is illustrative to visualize the resulting diverse deployment scenarios. In Figure 3-20 a reasonably complex example is given. For the three shown UEs a variety of user plane tunnels are needed. On the other hand, this complexity enables the optimal support for the different UEs with different needs for connectivity. The assumption is that all three shown UEs are subscribers of HPLMN A, thus roaming users in VPLMN B.

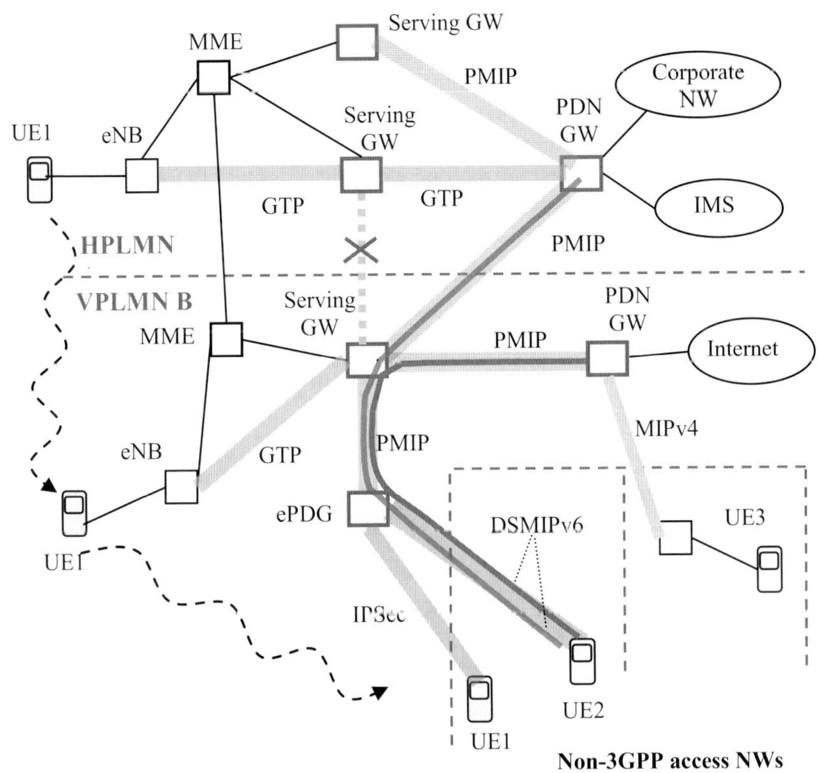

Figure 3-20: example 1 of deployment/operation scenario

For this example we note the following:

1. HPLMN A has started SAE with GTP variant for S5 and is now under migration (Serving GW with PMIP connectivity is already in place). As a consequence, at least one PDN GW has to be upgraded to dual protocol support.

2. VPLMN B has only PMIP for S5 (network internal) and S8 (roaming interface). The roaming agreement between HPLMN A and VPLMN B is PMIP based.

3. The inter-PLMN handover of UE1 between HPLMN A and VPLMN B will be S1 based i.e. via the EPC. Both Serving GW and MME are changed during the handover, but a temporary tunnel for the purpose of forwarding data packets during the handover is most likely not established, due to the different EPC protocol variants. It would be established, though, in case of both PLMNs being GTP based.

4. UE2 uses DSMIPv6 in untrusted non-3GPP access. This means that two mobility tunnels are in place between ePDG and Serving GW; the advantage is that the UE can handle mobility on its own, and maintain its IP address even when it subsequently enters another non-3GPP access network without PMIPv6 support. Also two tunnels (a DSMIPv6 tunnel inside the IPSec tunnel) are established between the UE and ePDG. In contrast, UE1 relies on network based mobility (PMIPv6) and only a single tunnel is needed.

5. UE1 uses the chained scenario, with the benefit that the mobility with untrusted non-3GPP access networks can be handled locally. This requires concatenation of PMIP tunnels at ePDG and saves some latency in signaling. Note that the similar scenario in trusted non-3GPP access (e.g. UE3 with MIPv4) is not possible.

6. UE2 has two PDN connections and in parallel, to different PDN GWs; one of them is in HPLMN and the other one, for local breakout, in VPLMN.

7. The typical traffic/service scenario is shown, with IMS and corporate network access as home operator's service and Internet traffic breaking out in the visited operator's domain.

Further deployment/operational scenarios are given in Figure 3-21 with special emphasis on the legacy system.

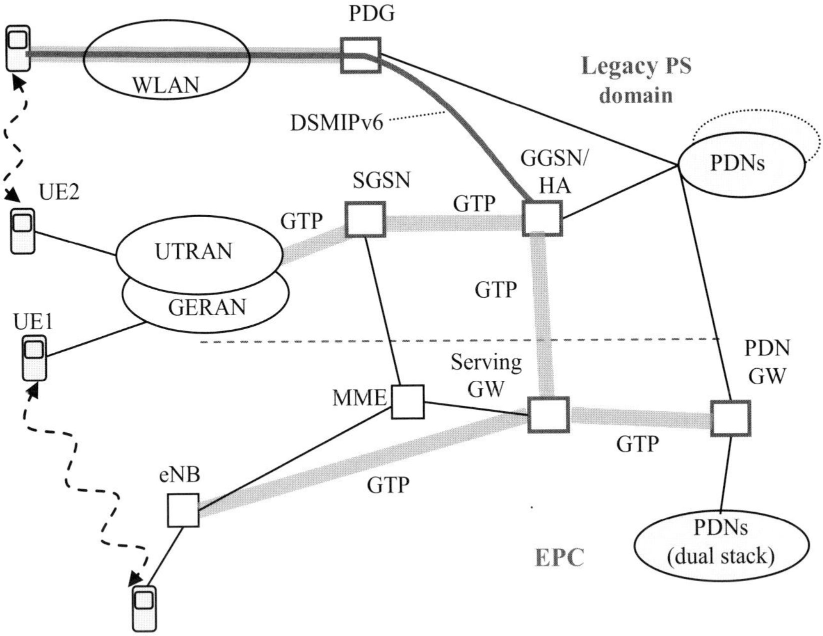

Figure 3-21: example 2 of deployment/operation scenario

Here the noteworthy observations are:

1. Most likely the network operator has just started to roll out E-UTRAN and EPC; this can be derived from the presence of a (pre-Rel. 8) PDG and an I-WLAN mobility service offering (indicated by HA, co-located with a GGSN).
2. UE2 is capable of I-WLAN and utilizes DSMIPv6 for basic mobility between GERAN/UTRAN (GPRS) and I-WLAN accesses. UE1 is at home (in DSMIPv6 sense) when in the GERAN/UTRAN access, and the DSMIPv6 tunnel is only necessary when in I-WLAN access. If the mobility is not needed, UE2 can avoid the tunnel overhead and let route the user plane directly from the PDG or GGSN towards the Internet.
3. UE1 moves between 2G/3G PS and E-UTRAN. A QoS mapping is done at handover between the two. In E-UTRAN, UE1 can benefit from dual stack bearers (i.e. transfer IPv4 and IPv6 packets on one bearer), whereas in GERAN/UTRAN it has to use single bearers (separate ones for IPv4 and IPv6 data packets).

3.13 Limitations of Rel. 8 and preview of Rel. 9/Rel. 10

These are the main and characteristic limitations of 3GPP Rel. 8:

- Only sequential utilization of multiple accesses is possible; in other words, if a new access is deemed more suitable than the current one, handover for all PDN connections is done together. For static conditions, multiple accesses can be employed, but only with multiple PDN connections. For Rel. 9 extensions are planned for allowing at least one non-3GPP access in parallel to a 3GPP access (for the same PDN connection!). This requires flow mobility, and solutions are proposed with DSMIPv6 flow bindings and equivalent means in PMIPv6.
- For GTP based EPC, no chaining is supported. This is due to issues with the concatenation of a GTP tunnel (in upstream direction) to a PMIP tunnel. The solution is targetted for Rel. 10.
- Multiple PDN connections to the same APN is supported only for GTP based EPC. In Rel. 9 a solution for PMIP will be sought.
- During the elaboration of the home cell concept after Rel. 8 (see Local IP Access topic in sub-section 3.10.2), operators showed a strong interest to flatten stil more the network architecture, including the macro cell part. Local breakout at eNodeBs is the target, reducing the load in the user plane (which is expected to rise dramatically, as soon as users turn heavily to broadband). This would be a major architectural change, and would have impacts on most key areas (mobility, charging, security).
- No optimized handovers (other than the one for CDMA © 2000 HRPD to E-UTRAN) are specified with non-3GPP accesses. The reasoning is that the potential non-3GPP radio technologies quite likely allow for dual radio operation, so that smooth handovers anyway can be achieved by the UE itself, if the new radio interface is activated early enough and the old one kept for a short time. This would then avoid complex handling in the network. Yet, this argument may not apply if the old radio access is lost very fast.
- No support for emergency communication (which may be a simple voice call or a multimedia session). Work for enhancement in this direction has already been started for rel. 9. It is currently foreseen that within the request for attach, the need for connection to an emergency PDN is indicated by an “emergency” type indication; the alternative solution, namely to let provide the UE an emergency APN was not pursued. For an emergency attach the security procedures need to be skipped, or reduced to a dummy behavior. A UE with ongoing emergency communication is restricted, e.g. it is not allowed to open an additional PDN connection or to request addi-

tional bearers (if these are needed for multimedia communication, they will be provided by the network).

- Service Specific Access Control (SSAC): due to the migration of all services to IP, the pure Domain Specific Access Control (DSAC) is no longer sufficient. (DSAC consists in selective barring of e.g. the CS domain during major disasters like earthquake or Tsunami for selected UEs, depending on their assignment to access classes, while allowing access in the PS domain.) With SSAC it will be possible to selectively throttle requests for Multimedia Telephony (MMTEL) services, i.e. either voice or video or both (see also sub-section 4.9.3).

Table 3-2 gives an overview of newly started work items in 3GPP for Rel. 9, which likely have a pronounced impact on several parts of the system. Release 9 shall be frozen in December 2009.

3GPP Rel. 9 Work Item	Impacts in ...
Enhanced eNodeB	architecture, physical layer, link layer, E-UTRAN, OAM
Support for IMS Emergency Calls over GPRS and EPS	NAS signaling (PDN connection, bearer handling, security handling)
SRVCC support for IMS Emergency Call	IMS message flows
Location Services for LTE and EPS	MME interfaces, additional EPC node (Evolved Serving Mobile Location Center)
Positioning in E-UTRAN	Radio protocol interfaces, S1, measurements in UE
MBMS support in LTE/EPS	Radio layer, S1, EPC interfaces, security, charging
Access Network Discovery and Selection Function enhancement	UE (combination of policies), roaming concept
Self Optimizing Networks	Enhanced self optimisation features in EUTRAN and OAM
Local Call Local Switch	Legacy CS radio access and core network.
HSPA enhancements	Further optimizations in the radio technology (MIMO, interference cancellation, dual cell)

Table 3-2: selected work items for 3GPP Release 9

Table 3-3 gives a listing of selected 3GPP's Rel. 10 work items; in some parts they encompass concrete specification, in other parts studies. Clearly this is less stable (when writing this text), and final decisions can only be taken at the time of freezing (stage 2 planned for Dec. 2009, and stage 3 for March 2010). But one aspect is definite: with Rel. 10 the next big step in radio technology is in progress – LTE advanced. It is related to the overall efforts of ITU-R's IMT Advanced and will boost the mobile data transmission rates to even higher values, entering officially the ground of 4G. The performance requirements are formulated in spectral efficiency (bits per second and Hz), so concrete values for user data rates depend on the employed bandwidth:

The minimum for peak spectral efficiency is 15 bits/s/Hz for downlink and 6.75 bits/s/Hz for uplink. At the example of a 40 MHz band the overall peak dat arate is 600 Mb/s in downlink and 270 Mb/s in uplink, whereas for a 100 MHz band the values are 1500 MB/s and 675 MB/s, respectively. The technical solutions are based on advanced MIMO techniques including, beamforming, carrier aggregation and relay/multipoint transmission. 3GPP has started a related study in [9].

LTE advanced
Local IP access
Multi-Access PDN connectivity and flow mobility
IMS based HomeNodeB
Local breakout and optimal routeing
GTP-based S8 chaining

Table 3-3: selected work items for 3GPP Release 10

References

[1] 3GPP TS 36.300: "Evolved Universal Terrestrial Radio Access (E-UTRA) and Evolved Universal Terrestrial Radio Access Network (E-UTRAN); Overall description"
[2] Volga Initiative, "VoLGA – Requirements V1.1.1 (2009-02)" and "V.o.L.G.A. Stage 2 V0.2.0 (2009-04-29)"
[3] 3GPP TS 23.216: "Single Radio Voice Call Continuity (SRVCC); Stage 2"
[4] IETF RFC 5213 (August 2008): "Proxy Mobile IPv6"
[5] IETF RFC 5555 (June 2009), "Mobile IPv6 support for dual stack Hosts and Routers"
[6] IETF RFC 3344 (August 2002): "IP Mobility Support for IPv4"
[7] OMA-ERELD-DM-V1_2: "Enabler Release Definition for OMA Device Management"
[8] IEEE 802.21, standard to be published (see http://www.ieee802.org/21/)
[9] 3GPP TR 36.913: "Requirements for Further Advancements for E-UTRA (LTE-Advanced)"

Chapter 4: Main Concepts

In this chapter we describe some fundamental concepts of LTE/SAE, elaborate on subsystems and functional components. We start with the radio access network and progress towards the core network, followed by additional network components (AAA infrastructure, policy and charging infrastructure, subscription and additional databases) and identity module in the terminal.

Obviously, the linear, sequential walk through these major components has some limitations; these have to play together, and the interfaces between them are common to the two corresponding sides. Some concepts anyway need to be spanned across several sub-parts of the system, and it is in the second part of this chapter where we elaborate on these (e.g. QoS, tracking, security, interworking in various variants).

4.1 E-UTRA Physical Layer

Radio resources (in one physical location) are in principle available only once in frequency and time; for multiple user access they have to be divided. The E-UTRA physical layer uses Orthogonal Frequency Division Multiplexing (OFDM) in **downlink** for multiple access on the radio resources. The same radio technology is used by IEEE 802.11 a/b/g (WiFi), IEEE 802.16 (WIMAX) and DVB/DAB. The concept is illustrated in Figure 4-1; it relies on multiplexing of channels in the frequency domain, where the carrier frequencies obey an orthogonality relation: if a signal is to be transmitted in a finite time/frequency rectangle, the magnitude of participating waves has zeros at several frequencies. With OFDM, the carrier frequencies are placed so that the fall into the zero points of magnitude of their neighbours, hence they are orthogonal. In this way approximately only half of the frequency band is required for a given number of carriers, compared to standard FDM, resulting in a more efficient transmission.

The basic element of OFDM transmission resources in E-UTRAN is given by one timeslot of 0,5 ms duration, either with 12 carriers of 15 KHz width, or 24 carriers of 7,5 KHz width (resulting in 180 KHz bandwidth, and 12 x 6 net OFDM symbols). A prefix is added before the OFDM symbol, to account for spreading of signal due to delays and guarantee synchronization. On the next level of granularity, various radio frames are defined for FDD (20 slots=10ms) and TDD (16 slots plus variable fields, in total also 10 ms).

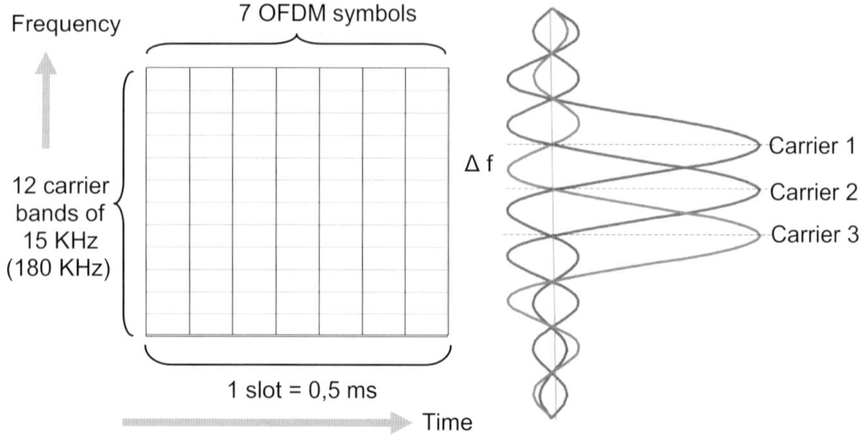

Figure 4-1: basic principle of OFDM in E-UTRA/LTE

The disadvantage of OFDM is the high ratio between peak and average power; it requires highly linear amplifiers with sub-optimal power efficiency. This would not be admissible for the mobile terminal equipment, and so a Single Carrier Frequency Division Multiple Access method (SC-FDMA) is used in uplink.

In order to enable high transmission rates, Multiple Input/Output (MIMO, actually multiplicity of antennas up to 4 for exploitation of spatial multiplexing), is supported.

The duplexing mode, that is, the allocation of frequency for uplink and downlink, can be by Frequency Division Duplexing (FDD) or Time Division Duplexing (TDD). FDD requires paired bands and was e.g. used in GSM (with equal width for uplink and downlink, according to the characteristics of the telephony service); TDD is suitable for an unpaired band. In E-UTRA/LTE most diverse radio band conditions are supported; up to ten FDD band configurations and 4 TDD bands have been defined (see Table 4-1), but their usage depends on national regulation; from the overall band(s) defined for LTE every operator has to acquire a license for a portion which should correlate with the expected subscriber base. For operators of legacy systems it is also attractive to have an option for "refarming", that is re-use GSM/UMTS frequency bands for LTE (but of course it would also depend on regulation).

FDD Band	Frequency uplink	Frequency downlink
I	1920 – 1980 MHz	2120 – 2170 MHz
II	1950 – 1910 MHz	1930 – 1990 MHz
III	1710 – 1785 MHz	1805 – 1880 MHz
IV	1710 – 1755 MHz	2110 – 2155 MHz
V	824 – 849 MHz	869 – 894 MHz
VI	830 – 840 MHz	875 – 885 MHz
VII	2500 – 2570 MHz	2620 – 2690 MHz
VIII	880 – 915 MHz	925 – 960 MHz
IX	1749,9 – 1784,9 MHz	144,9 – 1879,9 MHz
X	1710 – 1770 MHz	2110 – 2170 MHz

TDD Band	Frequency (up- and downlink)
a	1900 – 1920 MHz 2010 – 2025 MHz
b	1850 – 1910 MHz 1930 – 1990 MHz
c	1910 – 1930 MHz
d	2570 – 2620 MHz

Table 4-1: radio bands defined for E-UTRA

The physical layer fulfills these functions, and provides corresponding services to the higher layer:

- error detection on the transport channel and indication to higher layers;
- FEC encoding/decoding of the transport channel and mapping onto physical channels (including rate matching);
- Hybrid ARQ soft-combining;
- power weighting of physical channels;
- de/modulation of physical channels: QPSK, 16QAM and 64QAM are supported in downlink and uplink;
- frequency and time synchronization;
- radio characteristics measurements and indication to higher layers;
- Multiple Input Multiple Output (MIMO) antenna processing;
- transmit diversity (TX diversity);
- beamforming.

The physical radio resources are structured into physical channels according to Table 4-2.

Physical Channel	Physical Channel (long name)	Direction
PDSCH	Physical Downlink Shared Channel	downlink
PMCH	Physical Multicast Channel	downlink
PDCCH	Physical Downlink Control Channel	downlink
PBCH	Physical Broadcast Channel	downlink
PCFICH	Physical Control Format Indicator Channel	downlink
PHICH	Physical Hybrid ARQ Indicator Channel	downlink
PRACH	Physical Random Access Channel	uplink
PUSCH	Physical Uplink Shared Channel	uplink
PUCCH	Physical Uplink Control Channel	uplink

Table 4-2: physical (L1) channels

The main channel for data transport are PDSCH in downlink and PUSCH in uplink; PDSCH carries also paging messages. PDCCH transports control information e.g. on resource allocation. PHICH is used to transmit the feedback (acknowldegement) of a UE's data transmission on PUSCH. The PCFICH transport channel carries control format indicator information for the OFDM transmission. PMCH is foreseen for support of MBMS, and it relies on a "single frequency network" operation; in this mode multiple cells transmit in a tightly time-synchronized manner. PBCH is used for broadcast of essential information to all UEs in a cell (see sub-section 4.2.5.3). PRACH supports the synchronization of a UE with the network in order to prepare it for uplink transmission.

For further, more detailed descriptions of physical radio link properties (synchronization, modulation, channel coding, scrambling, measurements and power control), we refer the reader to dedicated sources in 3GPP TS 36.21x specifications ([1], [2], [3] and [4]), as well as to books more specialized on the LTE radio part (e.g. [5] and [6]).

4.2 E-UTRA Radio Link Layer and Radio Resource Management

4.2.1 Overview of Functions

Figure 4-2 gives an overview of functions in eNodeB and UE which together realize the Layer 2 and Layer 3 connectivity.

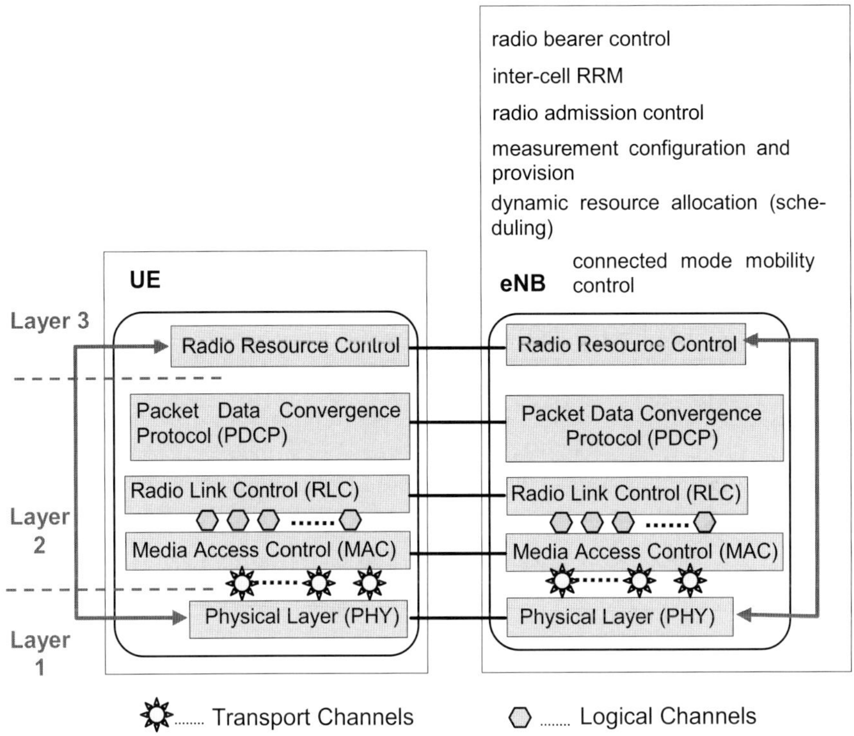

Figure 4-2: Functions in E-UTRAN and layered structure of Radio Link

A substructuring of layer 2, characteristic for a mobile network, is visible; further the concept of logical and transport channels is defined at interfaces (L2 internally and towards the physical layer). Figure 4-3 gives a more complete view of the embedding of the radio protocols used between the UE and the network; it makes apparent the split into control and user plane on the UE-eNodeB interface and shows also the higher layers. The layers discussed further in this section are shaded.

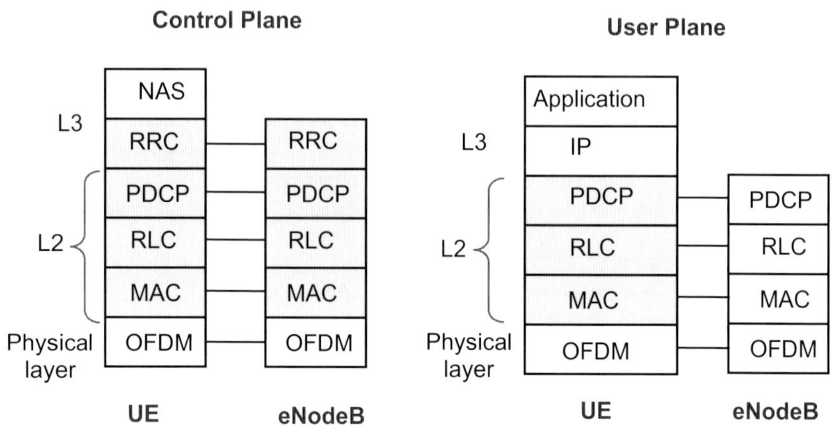

Figure 4-3: Protocol stack on LTE-Uu interface

☞ The LTE link layer is specified in 3GPP TS 36.321 [7] (MAC), TS 36.322 [8] (RLC), TS 36.323 [9] (PDCP). RRC for LTE is specified in TS 36.331 [10].

4.2.2 Logical Channel Structure

On layer 2 of E-UTRA several logical channels are defined, which map to transport channels as seen in Table 4-3. The further mapping of transport channels onto physical channels is then as follows:

in downlink: PCH and DL-SCH are mapped onto PDSCH, BCH onto PBCH and MCH onto PMCH;

in uplink: RACH is mapped onto PRACH, and UL-SCH onto PUSCH.

Logical Channel	Transport Channel	Direction
CCCH (Common Control Channel)	UL-SCH (Uplink Shared Channel)	uplink
DCCH (Dedicated Control Channel)		
DTCH (Dedicated Traffic Channel)		
BCCH *) (Broadcast Control Channel)	DL-SCH (Downlink Shared Channel)	downlink
CCCH (Common Control Channel)		
DCCH (Dedicated Control Channel)		
DTCH (Dedicated Traffic Channel)		
BCCH *) (Broadcast Control Channel)	BCH (Broadcast Channel)	downlink

PCCH (Paging Control Control Channel)	PCH (Paging Channel)	downlink
CCCH (Common Control Channel)	UL-SCH	uplink
DCCH (Dedicated Control Channel)		uplink
DTCH (Dedicated Traffic Channel)		uplink

*) BCCH can be mapped either onto DL-SCH or BCH

Table 4-3: logical (L2) channels and mapping onto transport channels

4.2.3 Layer 2 processing in downlink

A more detailed view of E-UTRAN's L2 processing in downlink is given inFigure 4-4. A clear layering into Media Access Control (MAC), Radio Link Control (RLC) and Packet Data Convergence Protocol (PDCP) is visible. The thick vertical lines indicate how data packets (more precisely, Service Data Units) pass the functional blocks and are processed sequentially or in parallel. (The direction is to be interpreted as downwards by the sending and upwards by the receiving side.)

PDCP contains link layer security (between UE and eNodeB) and header compression; the later uses the Robust Header (RoHC) compression scheme defined in IETF [11]. It is essential for IP transport over a radio link, which is subject to bandwidth constraints and transmission errors. Several different compression profiles, for the most heavily used combinations of network layer, transport layer and upper layer protocols (e.g. RTP over UDP over IP, TCP over IP, plain IP) are specified. The PDCP layer differentiates packet formats for (PDCP) user plane and (PDCP) control plane. The PDCP user plane contains 'real' (i.e. higher layer) user data or (higher layer) control plane data, whereas the PDCP control plane packets carry PDCP internal control information (status report, RoHC control data).

The Radio Link Control (RLC) layer is responsible for the actual data transmission across the radio link including supportive functions. Three transmission modes are available: acknowledged, unacknowledged and transparent.

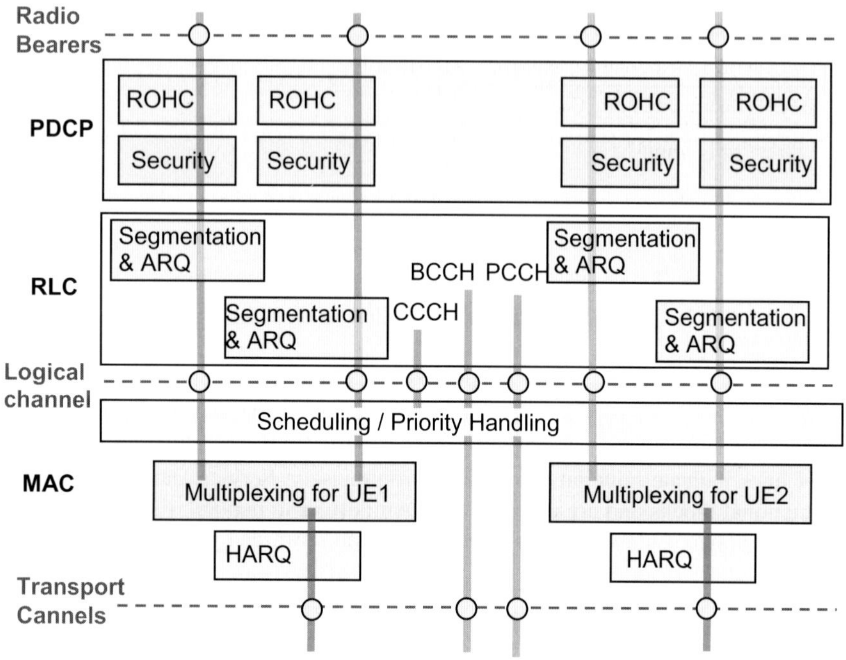

Figure 4-4: Layer 2 processing of LTE radio system in downlink

The transmission support functions differ for these modes:

- error correction through Automatic Repeat Requests (ARQ) for acknowledged transmission mode,
- concatenation/segmentation/reassembly, reordering and duplicate detection for acknowledged and unacknowledged mode,
- re-establishment of RLC by order of RRC,
- discard of data by order of PDCP.

Control channels (CCCH, BCCH and PCCH) act directly on the RLC and are thus exempt from processing by some functional blocks.

The MAC layer performs the mapping between logical channels and transport channels (including multiplexing and demultiplexing of radio frames, for all parallel bearers of all UEs served by this eNodeB). It handles the buffering and retransmission due to link errors. The Hybrid Automatic Repeat Request scheme (HARQ) is employed for the latter. Also the priority handling between UEs by means of dynamic scheduling (including reporting of scheduling) is task of MAC.

4.2.4 Layer 2 processing in uplink

Figure 4-5 shows the layer 2 processing of UE's data packets in uplink. Data streams are going through the functional blocks in PDCP, RLC and scheduling in MAC for radio bearers separately, and ar then multiplexed by the MAC layer onto a transport channel.

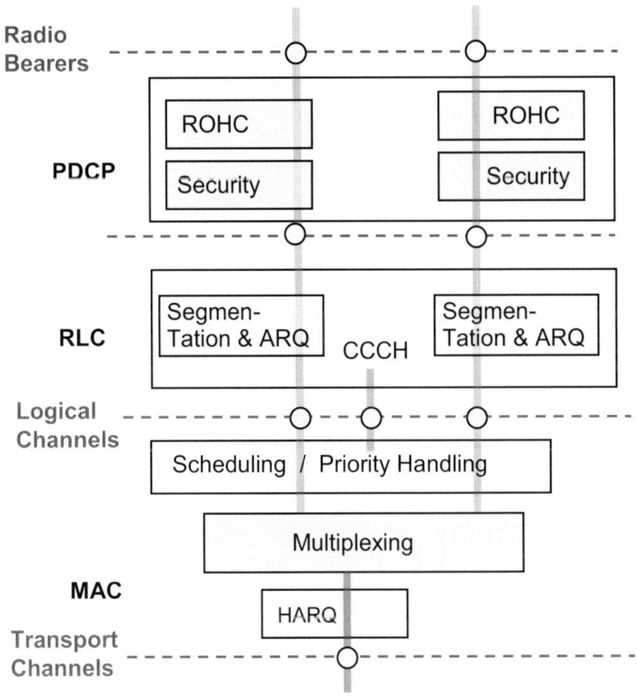

Figure 4-5: Layer 2 structure of LTE radio system in uplink

In uplink direction MAC defines also the random access procedure including the contention resolution (for the case that more than one request for access in up-link occur at exactly the same time). Another task of MAC primarily relevant for the UE is to maintain timing alignment for uplink transmission.

4.2.5 Radio Resource Control (RRC)

4.2.5.1 RRC States and RRC Connection Establishment

A state model for the resources on the radio channel is defined. A UE is in state RRC_CONNECTED if a radio resource context has been established with the eNodeB by RRC signaling. In this state data can readily be transmitted to the UE, and network controlled handover procedures, in particular for intra-RAT and inter-RAT mobility, are performed. The UE also monitors the paging channel and/or the System Information Block 1 (to detect change of System Information and potentially warning messages); it monitors the control channels to determine if it has data scheduled for reception. Further, it reports neighbour cell measurements and feedback on channel channel quality to the network.

The other state of the UE, regarding RRC, is RRC_IDLE; in this state the UE monitors the paging channel (to detect incoming connection requests, and as in RRC_CONNECTED, change of System Information and potentially warning messages). It also performs cell (re)selection.

In both RRC states the UE performs neighbour cell measurements, and acquires the System Information according to detailed rules.

4.2.5.2 Radio Bearers

On RRC level the radio bearers are controlled and the principle is sketched in Figure 4-6; radio bearers are the logical, resource controled "pipes" for transmission of data on L3 and higher between the UE and the eNodeB; a distinction is made between signaling radio bearers (SRB) and data radio bearers (DRB). The former ones are used for RRC messages and NAS messages; the latter are for user data. SRB0 is for initial RRC messages, and these are sent on the CCCH. SRB1 is for RRC messages and NAS messages, before SRB2 is established; the DCCH is used for their transmission. SRB1 and SRB2 are dedicated resources for a UE and thus need explicit establishment, like DRBs. An RRC connection requires the establishment of a SRB1.

In order to switch over from RRC_IDLE to RRC_CONNECTED the RRC connection is established via a three-way handshake (RRC CONNECTION REQUEST from UE to the eNodeB / RRC CONNECTION SETUP from eNodeB to the UE / RRC CONNECTION SETUP COMPLETE from UE to eNodeB).

SRB0 has no security protection; SRB1 and SRB2 are in principle integrity protected and ciphered (also termed encrypted), but there is the exception of the initial phase of dedicated RRC signaling, where the UE context has not yet been received by the eNodeB from EPC, and the first RRC message exchange on SRB1 is not security protected. SRB2 and DRBs are always security protected from the start (but DRBs are only ciphered, and not integrity protected).

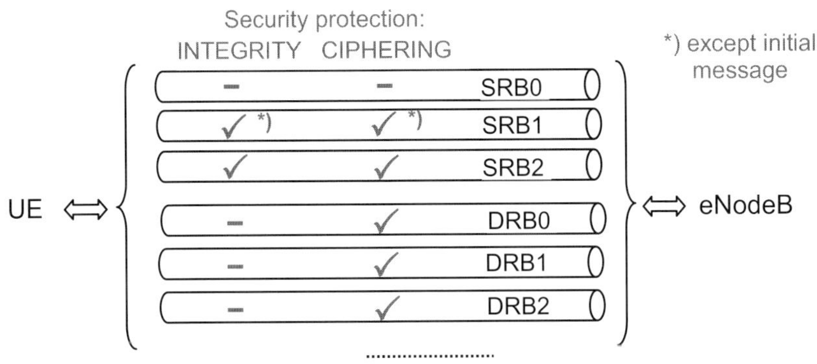

Figure 4-6: radio bearers between UE and eNodeB

The set of data used for security between UE and eNodeB is part of the Access Stratum (AS) configuration: algorithms for integrity protection, ciphering, key change indicator and a hop count; the latter two are used for security key determination at re-establishment of RRC connection and at handover.

Integrity and ciphering algorithms are always the same across SRB/DRBs, and are activated together; three different keys are used within the algorithms and derived from the base security key on eNodeB (K_{eNB}) (see also sub-section 4.8): one for the integrity protection of SRBs (K_{RRCint}), one for ciphering of SRBs (K_{RRCenc}) and one for ciphering of DRBs (K_{Upenc}). The security algorithms can only change at handover, whereas the four keys are changed at every handover and at re-establishment of RRC connection.

4.2.5.3 System Information

System Information (SI) is broadcast permanently by the eNodeB on the broadcast channel; it contains information relevant for all UEs in the cell, like physical data, cell access and selection parameters, type of cell, warning information. It is structured into System Information Blocks (SIBs, see Table 4-4). Not all SIBs need to be present; SIB Type 1 contains scheduling information about the System Information itself, so that quick access is possible.

SIB	Content: information on ...
Master Inform. Block	essentials on physical layer characteristics of the cell
SIB Type 1	conditions if a UE is allowed to access a cell, and scheduling of other system information blocks
SIB Type 2	common and shared channel

SIB Type 3	cell re-selection, mainly related to the serving cell
SIB Type 4	serving frequency and intra-frequency of neighbouring cells
SIB Type 5	E-UTRA frequencies and inter-frequency of neighbouring cells
SIB Type 6	UTRA frequencies and UTRA neighbouring cells
SIB Type 7	GERAN frequencies
SIB Type 8	CDMA2000 frequencies and CDMA2000 neighbouring cells
SIB Type 9	home eNodeB identifier (HNBID)
SIB Type 10	ETWS primary notification
SIB Type 11	ETWS secondary notification

Table 4-4: System Information (structure and content)

The network sends an indication on the paging channel, if System Information has changed; in this case the UE must read all System Information anew.

4.2.5.4 Radio Measurements

Both the physical and the link layer perform measurements. There are intra-frequency (at the same frequency as currently used by the UE) and inter-frequency, both as intra-RAT i.e. for E-UTRA, and inter-RAT (for UTRAN, GERAN, CDMA © 2000 HRPD and CDMA © 2000 1xRTT) measurements. They are performed by the UE on the downlink radio signal in connected mode according to the configuration provided by the eNodeB. The configuration describes the target frequencies and cells, reporting trigger, format and period. In addition to the cells listed in the measurement configuration provided by the eNodeB, the UE may, depending on the radio access, also report on detected cells. The signal reported back to the eNodeB is filtered (composed out of the newly measured value, plus a fraction of the previously reported signal) and thus smoothed out.

The reporting triggers are one of these:

- serving cell becomes better than threshold;
- serving cell becomes worse than threshold;
- neighbour cell becomes offset better than serving cell;
- neighbour cell becomes better than threshold;
- serving cell becomes worse than threshold 1 and neighbour cell becomes better than threshold 2;
- inter RAT neighbour cell becomes better than threshold;
- serving cell becomes worse than threshold 1 and inter RAT neighbour cell becomes better than threshold 2.

On the network side, the eNodeB also performs measurements; on the physical layer these are its downlink reference signal transmission power, received uplink interference power and uplink received thermal noise power. On layer 2 it is resource usage, number of active UEs per QCI and direction, packet delay and data loss. This information can be used for load balancing between cells and OAM monitoring.

4.2.5.5 Handover

If a UE in active mode needs to change the radio cell (mostly due to mobility, but could be due to cell overload), a handover is performed.

For intra-RAT handover, the network (the current or "source" eNodeB) decides if and when a handover is necessary, based on measurements, and prepares the target cell/eNodeB. The final command for handover execution comes from the target eNodeB and is relayed to the UE via the source eNodeB. After that, the UE accesses the target cell via the radio link and confirms the completion of the handover.

For intra-E-UTRAN handover there are two variants: within E-UTRAN, directly between source and target eNodeB (X2-handover, see sub-section 5.8.2), and indirectly, via the EPC (S1-handover, see sub-section 5.8.3). Also, there is the option to forward data received at the source eNodeB to the target eNodeB during the handover, in order to minimize data loss.

RRC in LTE defines also procedures for inter-RAT handover, for inbound/outbound handover from/to E-UTRAN to GERAN, UTRAN or CDMA 2000 ©. They differ slightly, due to the difference in interworking and integration between the corresponding systems, but details cannot be elaborated here and the reader is referred to the RRC specification [10].

4.2.5.6 Paging

Although paging is considered as part of RRC, it is discussed here separately (due to the prominent role and interplay with core network procedures).

The original purpose of paging in a mobile network is to search for the UE in idle mode, if the network has to send information (e.g. to start signaling for an incoming call); this is due to the fact that it can move freely within a defined area for which it is registered. In E-UTRAN it is a list of Tracking Areas (see subsection 4.5 for the underlying concept). Two other triggers for paging have been added in E-UTRAN, for UE in idle or in active mode:

- change of System Information, and
- arrival of an ETWS primary or secondary warning message.

The paging consists in transmitting indications on the PDCCH channel for paging groups (UEs belong uniquely to those groups), and on the PCH for a list of individual identities of UEs in the Paging Record. If a UE's identity matches the entries in the Paging record, it has to act according to the type of the page (either respond with NAS signaling message SERVICE REQUEST, to re-read the System Information, or to handle the ETWS warning messages, if it is capable of ETWS). Figure 4-7 sketches the actions upon reception of a paging message.

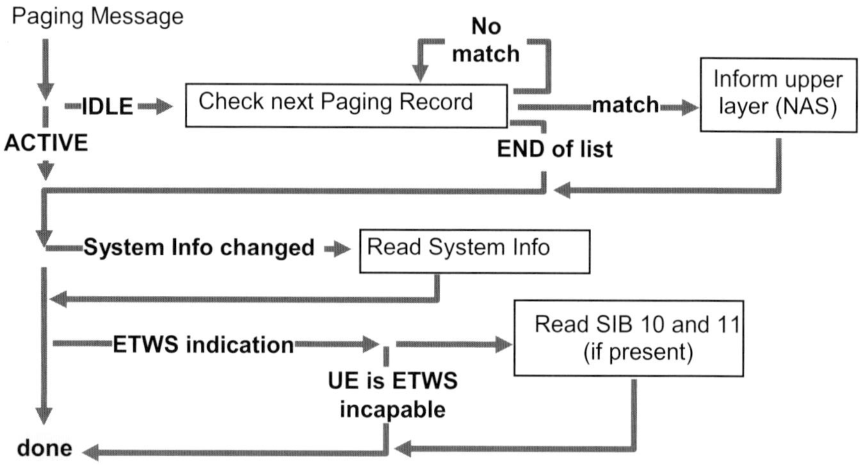

Figure 4-7: UE actions after reception of paging message

UEs are prepared to listen at their defined "Paging Occasions", which are calculated from their IMSI; there is also one predefined Paging Occasion for UEs without IMSI (as it is the case when a UE is operated without USIM). Discontinuous Reception (DRX) is an optimization for the purpose of battery saving, by which the UE can reduce the number of Paging Occasions to a minimum.

4.3 Evolved Radio Access Network

4.3.1 General

The Evolved UMTS Radio Access Network (E-UTRAN) is simplified to a large extent, as seen in Figure 4-8, compared to its GSM/GPRS and UMTS counterparts (GERAN and UTRAN, respectively).

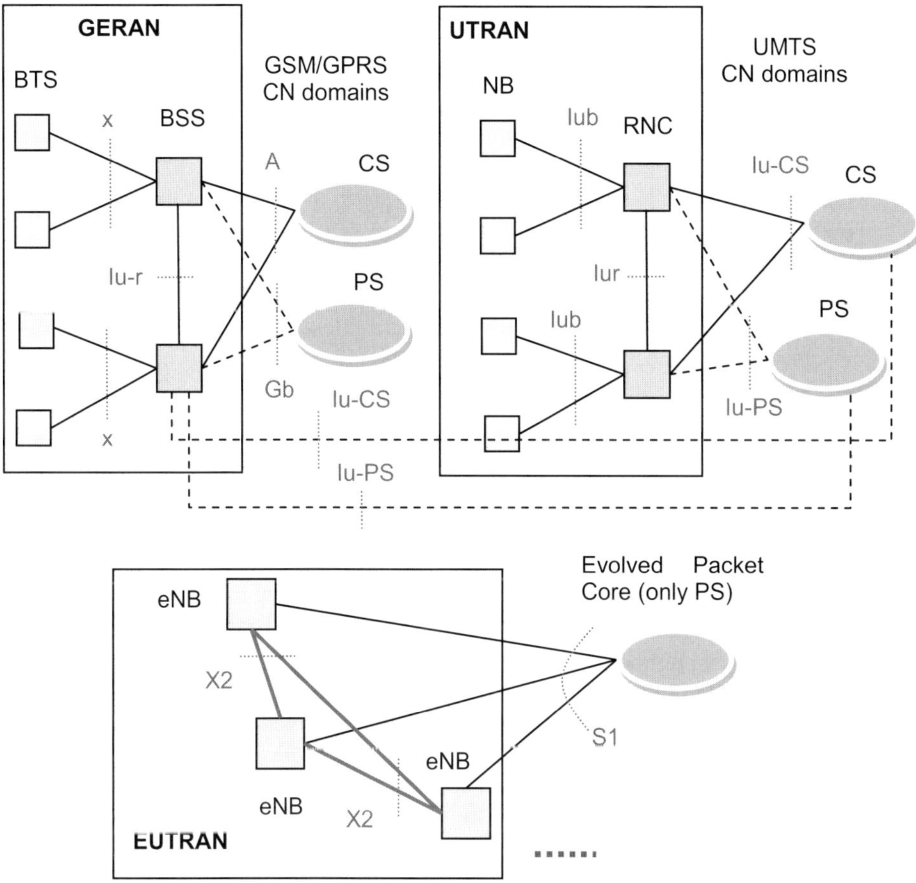

Figure 4-8: structure of radio access network for GERAN (upper left), UTRAN (upper right) and E-UTRAN (lower part)

For comparison we show the overall architectures of the 3GPP legacy systems. GERAN consists of the (radio) Base Station (BS) and Base Transceiver Station (BTS), see the upper left part in Figure 4-8. In UTRAN the constituting nodes are the radio (base) node (NodeB) and Radio Network Controller (RNC), as shown by the upper right part in Figure 4-8. Also shown is their interconnection, when GERAN BSSs are connected to the UMTS CN. Please note that both architectures are rather simplified (e.g. the broadcast domain is not shown).

☞ The reference architecture for GSM and UMTS networks is found in TS 24.002. E-UTRAN architecture is specified in TS 36.401 and UTRAN in TS 25.401.

In E-UTRAN the only node type is the Evolved NodeB (eNodeB), see the lower part of Figure 4-8. It connects to the EPC via S1 and to other eNodeBs via X2. Here only entities and interfaces of the radio network layer are shown, the transport network is not visible and is generally based on IP technology. Transport connectivity of eNodeBs with EPC nodes may be full or partial. In a larger network it would not make sense to connect eNodeBs in the eastern part of the network with Serving GWs in the western part; however, it may be more useful to connect them with remote MMEs.

Above the pure transport, S1 connectivity between eNodeBs and MME may be multiple; therefore the S1 sometimes is called a "flex" interface (the corresponding concept of connecting a radio access nodes to multiple core network nodes was called "Iu-flex" and not included in the original design of the 3GPP system, but introduced only in Rel. 5).

4.3.2 S1 Interface

This interface connects an eNodeB with two nodes in the EPC, MME and Serving GW, for the control and user plane, respectively.

Before presenting the details, it is worthwhile to touch a fundamental aspect in networking (and thus very relevant to mobile networks): the total communication can always be differentiated into real payload (also termed "user" traffic, because this is what a user has in mind and is willing to pay for) and signaling, which is needed to prepare, maintain and clean up the communication of payload channels. On the connecting interfaces one can most easily see the split (cf. dedicated bits of ISDN frames for B- and D-channel), but it is still an additional question if, in terms of network entities, separate nodes are foreseen for the two planes, or if the two logical planes are handled by one and the same node.

S1 thus also splits quite logically into user plane and control plane, and in this case separate entities are handling these "planes". The destination endpoints of the two protocol stacks on the network side are different, MME for control and Serving GW for user plane. In order to see the full picture, one should see the control plane relayed via MME to the Serving GW (see Figure 4-9).

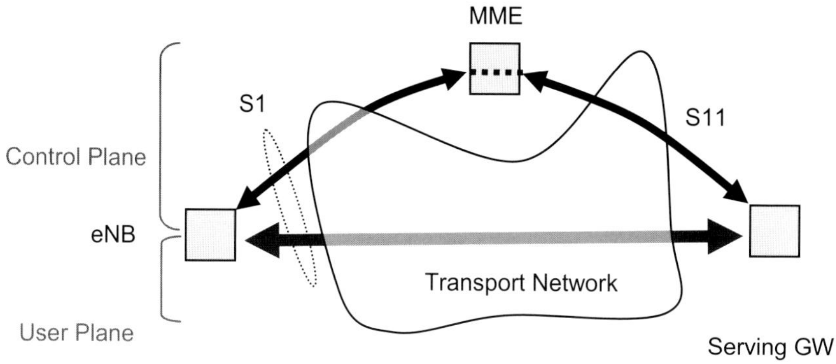

Figure 4-9: S1 control plane and user plane

Actually there was an intensive debate in the design phase on the advantages and disadvantages of such a split; at the end, the control plane has to "control" the user plane, so they must interact intensively. It is illustrative to compare the principle architecture of Figure 4-9 with the legacy CS and PS architectures of 3GPP: in the CS domain we see a combined core network node (MSC) from early GSM up to Rel. 99. In Rel. 4 the split was actually enabled, by defining MSC Server and MSC gateway, with a H.238 based interface between them. In the legacy PS domain exclusively the combined architectural approach was chosen (SGSN containing control and user plane functions). With the LTE/EPC architecture the split is introduced also here.

Figure 4-10 shows in a more detailed view the protocol stack for the S1interfaces. Both are based on (any) lower layers and IP, with another protocol on top; up to here this constitutes the transport network for the actually required S1 protocol functions. These latter are called S1-AP (S1 Application Protocol).

☞ S1 is described on the overall level in 3GPP TS 25.410; S1-AP is defined in 3GPP TS 25.413.

Above IP, the control plane uses SCTP (Stream Control Transmission Protocol), which can be characterized as a modern and flexible protocol for reliable transmission. The main protocol functions are:

- management (i.e. setup, modification and release) of Evolved Radio Access Bearers (E-RABs);
- context management (i.e. setup, modification and release) of contexts in the eNodeB;
- security (key handling for ciphering and integrity protection of the radio interface);
- handover signaling;

- paging (for finding the UE and establishing signaling, if it is camping on the eNodeB in idle mode);
- transport of NAS signaling messages;
- management procedures (for S1 itself): initial setup, reset, error indication, overload indication, load balancing, configuration update between eNodeB and MME;
- support of tracing (in connected mode);
- CDMA2000 tunneling;
- UE radio capability indication (to MME);
- location reporting: upon request of the EPC or when a change of cell occurs;
- warning message delivery: start and overwrite the broadcasting of warning message received from a warning center via the EPC;
- transparent transfer of RAN related information (between eNodeBs via the EPC).

For the user plane an encapsulation protocol was needed; luckily, it existed already in legacy 3GPP networks and it has proved extensively its suitability and efficiency: the GPRS Tunneling Protocol (user plane part) in version 2 and is run on top of UDP. For features and protocol structure see sub-section 6.1.3.

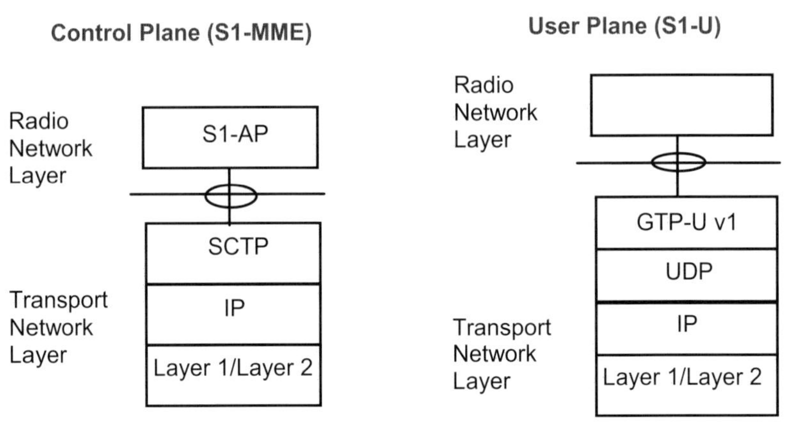

Figure 4-10: S1 protocol stacks for control and user plane

Both IPv6 and IPv4 are options for S1-MME and S1-U interfaces; DiffServ marking, as a local and simple QoS means (priorization), is mandatory for both interfaces.

The SCTP protocol for the control plane is especially suited for signaling (e.g. it has support for redundancy and in-sequence delivery); the SCTP connection is always established by the eNodeB as part of its initialization.

4.3.3 X2 Interface

This point-to-point, inter-eNodeB interface is mainly used during handovers. The protocol stack is an exact mirror of S1 and thus not shown by an extra figure: the control plane application is called X2AP (X2 Application Protocol) and is based on SCTP over IP over any L2/L1; similarly the user plane utilizes GTP-U over UDP over IP over any layer 2 / layer 1.

☞ The application part of X2 is described in 3GPP TS 36.423.

In more detail X2 has the following functions:

- support for intra-LTE mobility for UEs in connected mode, consisting in context transfer from source eNodeB to target eNodeB, user plane transport between source eNodeB and target eNodeB for data forwarding, handover cancellation and UE context release in the source eNodeB;
- load management: neighbouring eNodeBs exchange data on their resources, traffic load and possibly overload;
- general X2 management and error handling functions (setup, error indication and reset);
- application level data exchange between eNodeBs;
- trace functions.

See sub-section 6.8 for an overview on X2 protocol messages.

4.4 MME Load balancing

Load balancing is achieved between MMEs within one pool area by setting a percentage value per MME and instructing the eNodeBs (via S1-AP signaling) to select the MMEs with a probability accordingly, for UEs entering the pool area. The concept is seen from Figure 4-11.

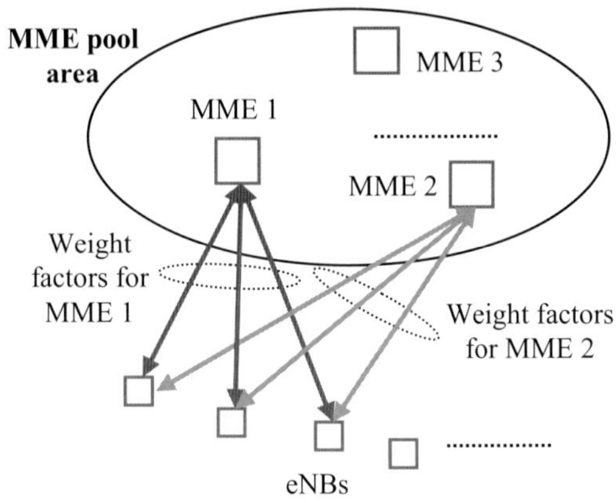

Figure 4-11: concept for MME load balancing

The percentage values (weight factors) would reflect the proportional capacity of MMEs of the total in the pool. This procedure has a slow dynamics; e.g. if a new MME is installed, its load would only go up gradually. Yet, in this case the operator can set the percentage initially to a much higher value than what is the "real" share of this MME, monitor the load carefully and decrease the percentage value subsequently.

Load re-balancing may be needed, e.g. if an MME needs to be taken out of service; in this case the percentage will be set to zero, but this affects only UEs arriving newly in the MME pool area. It would be too slow in practice. Therefore the MME, based on configuration, can release explicitly the S1 connection for UEs in active mode, setting the cause to "load balancing required" (see the sequence of steps [1] –[4] for UE1 in active mode in Figure 4-12, left part):

[1] MME1 as the "old" MME releases the S1 connection, indicating a cause of load re-balancing;

[2] the eNB currently handling the UE similarly releases the RRC connection with the same cause;

[3] the UE re-establishes the RRC connection, indicating back to the eNB the load rebalancing indication;

[4] the eNB selects MME2 as "new" MME for the UE.

In a certain time frame before e.g. an MME needs to be taken out of service, the MME can also instruct the UEs in idle mode requesting TAU or attach to do another TAU afterwards, with load balance indication (the steps are [a] – [f] for a UE2 in idle mode, visualized in Figure 4-12 on the right).

[a] UE2 performs a TAU or attach procedure to MME2 as its currently assigned ("old") MME;

[b] step [a] is accepted by MME2 (so it is not yet in overload);

[c] at some point in time later overload occurs in MME2 and is signaled to the eNB handling UE2;

[d] and

[e] are the same steps as [2] and [3] for UE1, i.e. release and re-establishment of the RRC connection with cause of overload;

[f] UE2 performs a TAU, which is routed by the eNB2 to an MME not o=under overload (in this case MME1 is assumed to be no longer in overload or out of scrvicc).

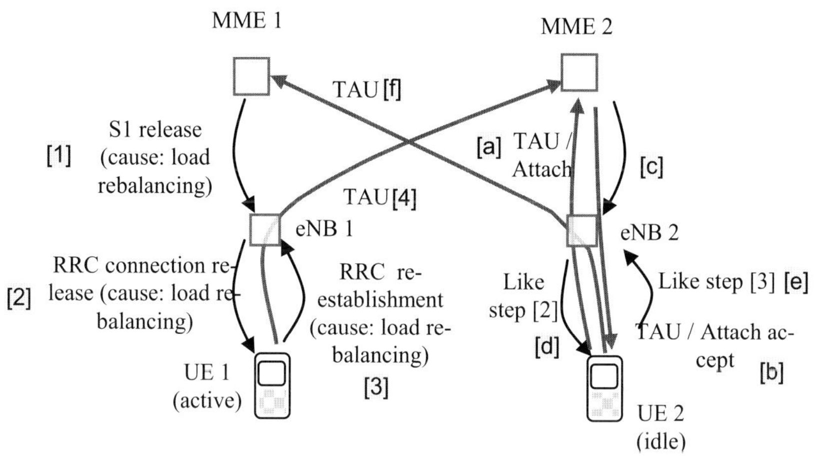

Figure 4-12: concept for MME re-balancing

Additionally, if an MME runs into high load, it can start the overload procedure; it consists of sending an OVERLOAD START via the S1-AP message to connected eNodeBs (at random). The eNodeB then rejects all or selected (e.g. mobility related) non-emergency RRC requests. When load has dropped sufficiently on the MME in question, it sends OVERLOAD STOP messages to the relevant e-NodeBs.

4.5 Tracking Concept

Clearly, in active mode the network has full knowledge on the UE's presence in the network (due to the necessity to manage the resources within corresponding cells).

In contrast, in idle mode, as long as the UE is principally "registered" with the network, it needs to know UE's location at least on the level of some area. If this was not the case, a search in the whole network would be required at the point when terminating traffic arrives. The concept of "Tracking Areas" (TAs) was developed for this purpose. It is similar to Location Areas in GSM and Routing Areas in GPRS.

Basically the whole E-UTRAN is divided into non-overlapping TAs (see Figure 4-13). Each TA is uniquely identified by a TA identification (TAI, see sub-section 4.5), which is broadcast per E-UTRAN cell (and every cell is uniquely assigned to only one TA). Note, however, that one eNodeB may have cells assigned to different Tracking Areas. The UE needs to register its currently seen TA with the MME, and MME allocates a list of Tas to the UE. The UE can move freely within these Tas, without any need to update its registration status; it simply compares the TA received in the broadcast from the currently camped-on cell with the list, and only if no match is found the UE needs to register again with MME. This mobility procedure is called "Tracking Area Update" (TAU, see sub-section 5.8.1). How the lists are created by the MME is not specified, and is thus left implementation dependent. The UE also has to re-register periodically (based on a timer configurable by the network operator).

The MME can update UE's TAI list in two NAS signaling procedures (attach procedure and TAU procedure); the TAI list is used by the MME in the paging procedure, to limit the distribution of the page (see sub-section 4.2.5.6). Neglecting abnormal cases, the MME can then be sure to find the UE by this limited page.

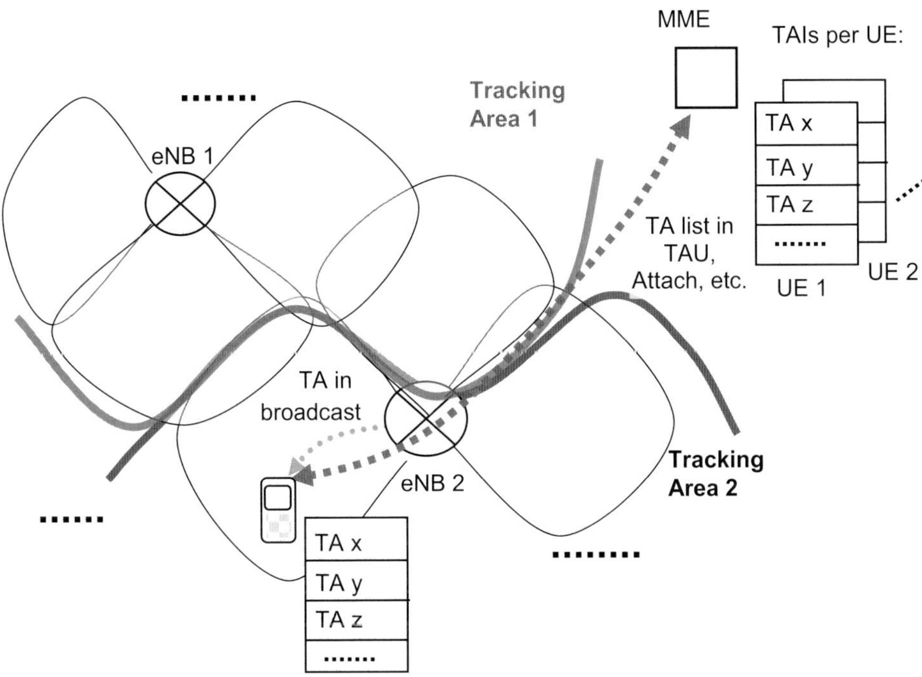

Figure 4-13: Tracking Area concept

Selected Tracking Areas may be forbidden individually per user (one for regional provision of service and one for roaming), based on operator's policy; this allows a fine-grained service and roaming control. For that purpose the UE has to store two lists, and it inserts TAIs into these lists, whenever a reject message with a corresponding cause is received from MME. The lists are cleared at switch-off of the UE and when the UICC (containing the USIM) is removed from the UE; on the other hand, TAI entries are removed from these lists if the MME allocates these TAIs in the above mentioned update procedures (obviously the previously imposed restrictions are then no longer valid).

4.6 IP Packet Bearer

The Evolved 3GPP system is based on packet switching, in particular on Internet Protocol, technology. Multiple services and connections may be handled simultaneously, and they may need differentiation in various respects, e.g. in their Quality of Service, charging characteristics or version of the provided IP connectivity. An underlying basic concept employed for this purpose is the (IP packet) "bearer" concept; a bearer is the finest level of granularity with respect to different handling of data packets transferred via the EPS. Bearers are associated with a PDN connection and with packet filters, to allow the differentiation between all packets for this PDN connection.

In comparison, the concept of "bearers" in CS technology is quite different (for GSM see 3GPP TS 22.001 [12] and 3GPP TS 22.002 [13]): there the QoS parameter set is very reduced (mostly fixed data transmission rate and delay with some error rate, or improved error rate with variable data rate and delay); other bearer characteristics include e.g. the capability of information (like speech or unrestricted digital information), access and interworking features of the information transfer.

It is important to note that the scope of the Evolved 3GPP system, and thus EPS bearer, can at maximum extend up to the point where the interconnection with external networks occurs, i.e. up to the Sgi reference point. But there is a difference for the two supported variants of S5/S8 interface:

1. In GTP based EPS the EPS bearer extends between UE and PDN GW (GTP has all built-in functionality for bearers).
2. In the PMIP based EPS architecture the EPS bearer extends between UE and SGW; between SGW and PDN GW there is only "pure" IP connectivity (as seen from the EPS mobility tunneling point of view; in the underlying IP transport network there may well be means for differentiation, e.g. typically DiffServ for the transport of IP packets). Independent of the lacking bearer capability between Serving GW and PDN-GW, also the signaling capability does not exist in PMIP (in the GTP case it is contained in GTP-C messages). This part can be added again by employing dynamic Policy and Charging (PCC), realizing another channel of control. Note, however, that dynamic PCC is optional, and its utilization in roaming is not a legacy mechanism; therefore some initial barrier in deployment/rollout can be expected.

Between UE and the SGi reference point we can subdivide the EPS bearer into portions, each one between particular nodes on the path: radio bearer between UE and eNodeB (see sub-section 4.5.5.2), S1 bearer between eNodeB and Serving GW, S5/S8 bearer between Serving GW and PDN GW; this is visualized in Figure 4-14, where the EPS bearer concept is also compared with the bearer model for the legacy PS domain and its interworking leg S4 between SGSN and Serving GW). For non-3GPP access there is no equivalent bearer model, at least it is not supported by the Evolved 3GPP system.

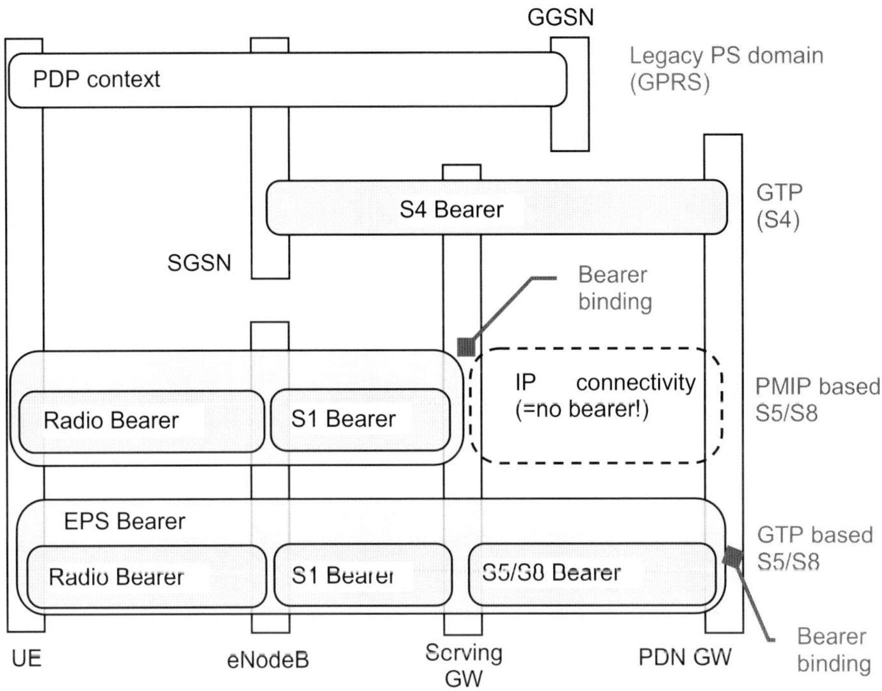

Figure 4-14: EPS bearer models, compared to PDP contexts in the legacy system

QoS for a bearer consists in the treatment of packets, like scheduling policy, queue management policy, rate shaping policy, radio link control. One EPS bearer is always established together with the connection to a PDN (the default bearer). Additional bearers can be added and removed by dedicated bearer handling procedures, invoked by NAS signaling between the UE and the network; these are called dedicated bearers.

A distinction is made between Guaranteed Bit Rate (GBRs) and non-GBR bearers; GBR bearers get resources permanently allocated (at the time of bearer establishment or modification). A default bearer is always a non-GBR bearer. Details of these QoS properties are presented in sub-section 4.11. Another distinctive feature is the IP version capability of a bearer; it may be either IPv4, or IPv6, or IPv4 and IPv6, i.e. dual stack. (Note: in the legacy system only single IP version bearers, i.e. IPv4 and IPv6 bearers are supported).

With multiple PDN connectivity, several bundles of bearers will exist; they may spread out from Serving GW, depending on whether connectivity to the different PDNs is enabled by one and the same or via multiple PDN GWs (see also sub-section 4.20). On the radio link and between eNodeB and Serving GW they are established in parallel (for the case of 3GPP access at least); gives an example of a bearer arrangement for three APNs, realized via three PDN connections (one per APN, consequently three default bearers), two PDN GWs and consisting of a total of five bearers. Note that:

- radio and S1 bearers may be temporarily inactive,
- PDN GWs may reside in HPLMN or VPLMN, and
- the UE will have to handle four IP addresses (one for APN 1 and 3, and two for APN2).

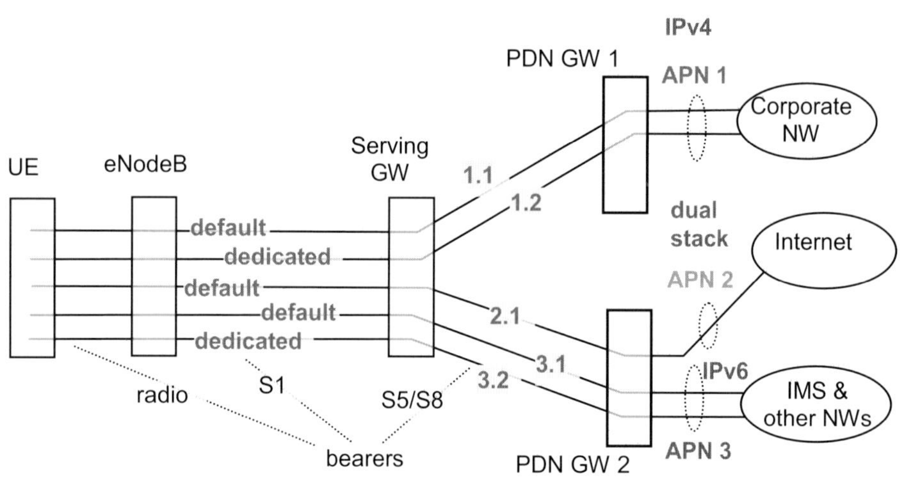

Figure 4-15: example arrangement of bearers (with multiple PDN connectivity)

In Rel. 8 only unicast bearers are defined; extensions for multicast bearers may be expected for future releases (e.g. for MBMS service in Rel. 9).

Related to the bearer concept are the concepts of Traffic Flow Template (TFT) and packet filters. A TFT consists of a set of packet filters and is used to map a traffic flow aggregate to a bearer, described by a quintuple of source and destination IP address and port, and protocol id. The default bearer has a match-all TFT. The detailed handling in network nodes and mapping between traffic flows, transport QoS on S5/S8 and EPS bearers is visualized, see Figure 4-16 for GTP on S5/S8:

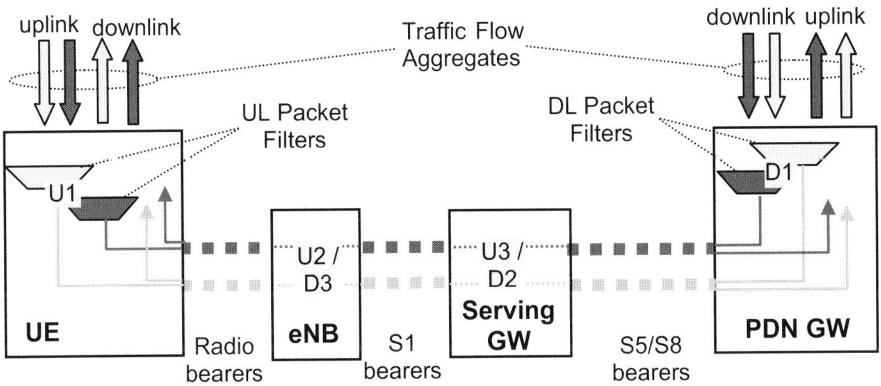

Figure 4-16: details of EPS bearer handling (GTP case)

For GTP as EPC protocol, mappings are done at the following points in uplink (in brackets the corresponding labels within Figure 4-16):

are given):

1. in UE (U1): according to uplink TFTs onto Radio Bearers;
2. in eNodeB (U2): one to one from Radio Bearers onto S1-bearers;
3. in Serving GW (U3): one to one from S1-bearers onto S5/S8 bearers.

And in downlink similarly:

1. in PDN GW (D1): according to downlink TFTs onto S5/S8 bearers;
2. in Serving GW (D2): one to one from S5/S8 bearers onto S1 bearers;
3. in eNodeB (D3): one to one from S1 bearers onto Radio Bearers.

As a characteristic difference with PMIP as EPC protocol, the mapping in PDN GW is instead according to downlink TFTs, and similarly in Serving GW from S1 bearers, onto Transport NW level QoS parameters (in practice DiffServ code points). Also, in Serving GW according to downlink TFTs onto S1 bearers.

In conclusion, packet classification for downlink traffic takes place: (1) once, at the PDN GW in case of GTP based EPC, and (2) twice, at both the PDN GW and

Serving GW in case of PMIP based EPC. For uplink traffic the packet classification happens in the UE for both EPC protocol variants. Down-link "bearer binding" takes place in PDN GW for GTP based EPC, and in Serving GW for PMIP-based EPC.

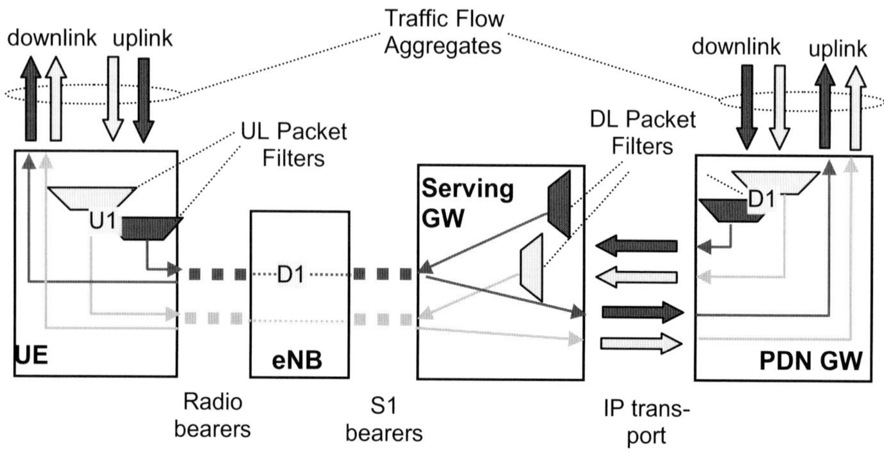

Figure 4-17: details of EPS bearer handling (PMIP case)

The classification scheme sequentially checks the match of a packet against a TFT, based on the priority of TFTs; if no match occurs for any TFT, the packet is assigned to a bearer which has no TFT assigned, or, if all bearers have TFTs assigned, it is discarded.

In case of PMIP as EPC protocol, the downlink packet classification within PDN GW can be used to enable DiffServ marking as a local QoS handling scheme on IP transport level. Yet, the fact that then **only** DiffServ marking of packets remains between Serving GW and PDN GW illustrates again the characteristic feature of the "bearer-less" PMIP protocol.

4.7 Overall concept for authentication, authorization and accounting (AAA)

AAA (Authentication, Authorization and Accounting) relates to the functions ensuring that the valuable resources of the operator's network are used only as they are allowed, and that this usage can be accounted for (either later in postprocessing including billing, or in real time).

A cornerstone of AAA for the basic mobile services in the 3GPP system is traditionally access authentication; any 3GPP access (whether in the HPLMN or VPLMN) is naturally considered as trusted, and a UE's security credentials are downloaded from the HSS to the core network node in charge of access authentication (MSC/VLR in legacy CS, SGSN in legacy PS, and MME in case of E-UTRAN). Authentication is required in order to guarantee that the requesting UE is linked to the identity it claims to be; authorization enforces that only those resources and services are used, for which the user has a subscription (and eventually pays for). Accounting is the function providing all necessary data to the operator for setting up the final bill and also collecting statistics, i.e. counting/metering the user traffic of different type and recording charging relevant events.

Whereas the legacy interfaces with HSS for AAA purposes (reference point D between MSC/VLR and HSS, and reference point Gr between SGSN and HSS) are using the MAP protocol (a member of the CS-type SS7 protocol family), the new system relies purely on the IP based DIAMETER protocol (for protocol details see sub-section 6.5), both for the 3GPP access (=E-UTRAN) and the non-3GPP access part. Note that DIAMETER and the equally IP based, but vastly different protocol RADIUS are already in use for I-WLAN AAA functionality. Also, the AAA server, originally put in place for I-WLAN, is reused and extended for non-3GPP access related AAA procedures; this includes also the interface between AAA server and HSS.

An overview of AAA interfaces, protocols used and functionality provided, is given in Figure 4-18 (for comparison also including the legacy nodes pre-Rel. 8 SGSN [relevant in the case of Gn/Gp interworking], and MSC/VLR [relevant e.g. if CS fallback is used]).

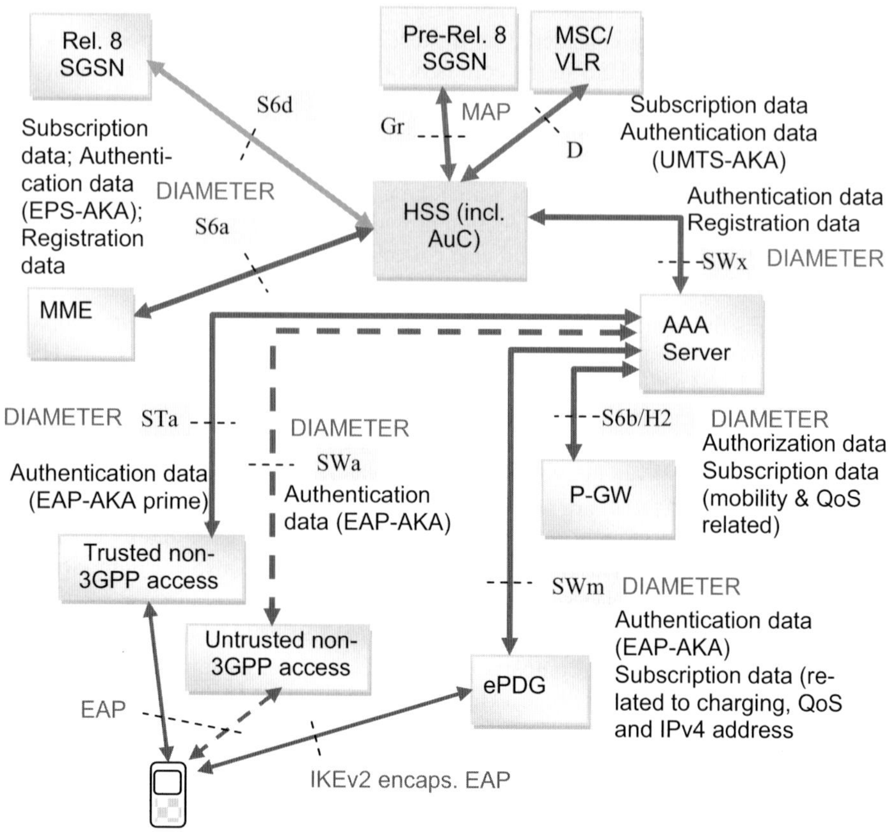

Figure 4-18: AAA interfaces, protocols and functionality

The central anchor point of AAA data and thus "center of trust" is HSS (here for simplicity shown as including the Authentication Center); it stores the long term credentials of users/UEs and generates the primary data sets for authentication (the so-called authentication vectors). Also, it contains the database for subscriptions; authentication and subscription data is delivered to the control nodes within the different domains (SGSN, MME, AAA server for propagation to non-3GPP access).

In 3GPP access via EPC the authentication data is used in the authentication procedure according to EPS-AKA. As an integral part of the security concept it is further discussed in the next sub-section.

In non-3GPP access there are four AAA signaling "channels":

1. STa between trusted non-3GPP access and AAA server: an authentication and authorization procedure is mandatory and it is based on EAP-AKA' (for clarity written here EAP-AKA **prime**). It is possible to separate the security domain per such trusted non-3GPP access by using the access network id (ANID) within its key derivation function, as far as different AN-IDs are used. EAP-AKA prime was specifically developed for this environment. STa is complemented by EAP transport over any L2 towards the UE.
2. SWa between untrusted non-3GPP access and AAA server: optionally an authentication procedure, based on EAP-AKA, can be used. It is not mandated, because in untrusted non-3GPP access anyway an authentication and authoriztion procedure for access to EPC is to be performed (see item 3 of this list). However, independent of access to EPC there is also a need for L2 security configuration, and it might be useful for operators to leverage the 3GPP AAA infrastructure also for this purpose within the non-3GPP access (e.g. providing key material for WLAN radio link). SWa is complemented by EAP transport over any L2 towards the UE.
3. SWm between ePDG and AAA server: this realizes the mandatory authentication and authorization in conjunction with IPSec tunnel setup between UE and ePDG. SWm is concatenated to IKEv2 signaling (which in turn encapsulates EAP) towards the UE.
4. S6b/H2 between PDN GW and AAA server: it is used to deliver general authorization data (e.g. default APN, authorized APNs, PDN type, static IP address), and authorized features of mobility with non-3GPP accesses (for all MIP variants within EPS). Additionally subscription data like type of charging and (static) QoS profile, as well as trace information data can be conveyed. Further, the PDN GW address itself must be registered with the AAA server. For the case of I-WLAN mobility architecture the reference point H2 applies with data/features reduced for DSMIPv6 only.

More information on the authentication part in message flow level is found in sub-section 5.1, and further details of AAA messages in an examplary manner in sub-section 6.5.

☞ AAA interfaces in EPS are specified in TS 29.273 [14].

4.8 Security concept

4.8.1 Basic security concept for E-UTRAN and EPC

Security is a key feature in any mobile network, and it contributes considerably to the complexity. The starting point of the security design for the Evolved System was the vast and positive experience with 3GPP's security concept from 2G and 3G. On the one hand this was to be leveraged, while it was also necessary to apply improvements and to account for new functionality. In the following we proceed in steps, explaining first the security concept for LTE access (i.e. when a UE connects via E-UTRAN) to the EPC, followed by security procedures applied when interworking with the legacy 3GPP occurs; finally we turn to security for interworking with non-3GPP accesses. Related procedures are explained in more detail in chapter 5.

The desired (and finally implemented) security features are similar to those in the legacy 3GPP network:

1. authentication: guarantee that the real identity of a UE is the one it is claiming to be;
2. user identity confidentiality: ensure that other parties than the network operator cannot determine a UE's permanent identity (IMSI, or derived from IMSI); this is achieved by the use of temporary identifiers as far as possible. However, in some situations when signaling to the network (e.g. after all context data had been erased) the permanent identiy needs to be used;
3. device confidentiality: ensure that other parties than the network operator cannot determine the identity of a user's terminal;
4. ciphering (=encryption) of signaling and user data: a transformation is applied to the original data using a security algorithm, making it non-understandable to parties without knowledge of the appropriate security credentials;
5. integrity protection of signaling and user data: a security algorithm is applied on the original (user or signaling) data, generating a unique code which depends strongly on the security credentials. Any change to the original data would lead to a vastly different code. In this way the data is protected against change at intermediaries.

The security concept is built on reuse of USIM, the UMTS subscriber identity module contained on an UICC. By its very nature this constitutes a security relation between the UE (into which the UICC is inserted) and the home 3GPP network. Further security relations and related security key data, for a visited 3GPP network and access networks, can be derived from it after performing a kind of bootstrapping procedure (authentication and key agreement – AKA). For access to EPS, the EPS-AKA procedure is specified (see sub-section 4.8.3); a security context created by this procedure is called "native".

A mapping from a UMTS security context onto an EPS security context is also possible; this allows a more seamless inter-operation with the legacy network (handover/intersystem change). A security context needs to taken into use explicitly, so a differentiation into current and non-current security context is made. A principal difference between EPS and 2G/3G, where only one security context can exist at a time.

The overall situation is as shown in Figure 4-19. We see the different access domains under the umbrella of the EPC, and basic components of security spanning across them. The UE contains the UMTS Subscriber Identity Module (USIM), holding the long-term credential shared with the HSS (here it is assumed, for simplicity, that the Authentication Center/AuC, in principle logically separate, is collocated with HSS). Security procedures are run, establishing security contexts (indicated in Figure 4-19 by pentagons) in access network and core network entities, as well as in the UE. Nodes in the 3GPP network domain are considered to be trusted and communication with them secure (in fact this is guaranteed from the overall 3GPP specifications for network domain security); authentication data can be distributed to them to let perform the authentication. In general, nodes in non-3GPP (access) networks cannot be considered trusted in 3GPP's sense, therefore authentication data cannot be delivered to them; they can only act as pass-through authenticators. Thus the authentication for non-3GPP accesses must always be performed with the AAA server in the home network; the method is based on the IETF defined EAP scheme, with a 3GPP defined application (AKA).

Due to the modular concept, the detailed security algorithms are in principle flexible and extensible; however, two integrity and two encryption algorithms are generally defined: the 128bit EEA1, based on SNOW 3G algorithm, and the 128bit EEA2 algorithm based on AES. Because potentially several algorithms might be supported, the UE and the network need to agree on the algorithms to be used; this is achieved by security mode commands. They exist for AS and NAS level. Ciphering is optional in the network, and for that purpose a "Null" algorithm for ciphering is also defined.

One of the fundamental design principles was not to let EPS security break down, even if a breach in security within a particular non-3GPP access would occur. This is achieved by including access network related information in the determination of security context data, which leads to a separation of security domains.

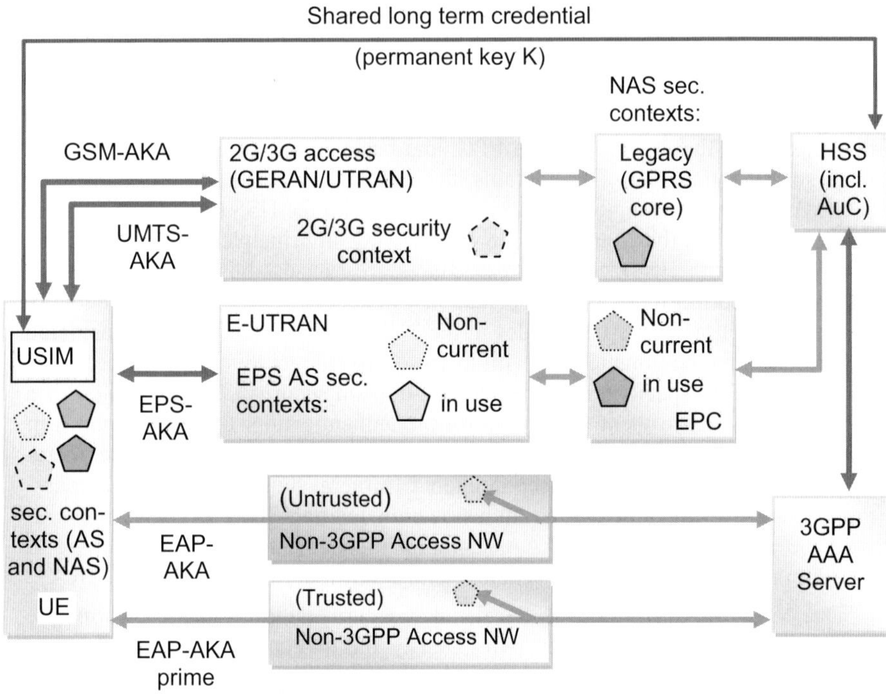

Figure 4-19: basic components of security of the Evolved 3GPP system

Note that the SIM (the 2G variant of a subscriber identity module) is not allowed for access to EPC; the motivation is that it is in principle weaker, and it was considered not to be appropriate to fall back to such a lower level of security after having raised it already with the 3G technology.

Due to the difference in architectures between legacy and Evolved 3GPP system also differences in the provisioning and usage of security keys arise; further, additional security mechanisms due to enhanced capabilities are needed (e.g. for protection of mobility signaling).

☞ The overall security concept for access to the EPC is defined in 3GPP TS 33.401 (for 3GPP access) and 3GPP TS 33.402 (for non-3GPP accesses).

4.8.2 EPS security key hierarchy

Security keys of different type need to be used in different parts of the system; Figure 4-20 gives the overview. It features as the starting point a permanent key, shared between USIM and AuC, and successive key derivations in a key hierarchy with distribution to various parts of the system. The EPS specific authentication and key agreement (EPS-AKA) procedure is used within this process (and is explained in the next sub-section). Compared to the legacy 3GPP system the key hierarchy is now more elaborate.

Note: ASME is a general term for "Access Security Management Entity", which in this context is identical to MME.

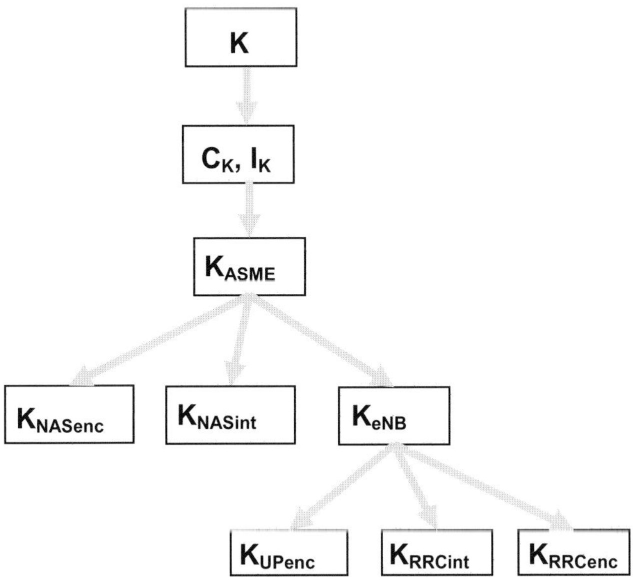

Figure 4-20: security key hierarchy in EPS

K: long term credential, shared between UE and AuC;

C_K: ciphering key: result of an EPS AKA run; shared between UE and HSS;

I_K: integrity key: result of an EPS AKA run; shared between UE and HSS;

K_{ASME}: key for access security management entity (corresponds to MME), derived from C_K, I_K by an EPS specific key derivation function; shared between UE and MME;

K_{eNB}: key for eNB security from which further E-UTRAN specific keys are derived; shared between UE and MME;

K_{NASenc}: key for NAS encryption in EPS, specific per selected encryption mechanism; shared between UE and MME;

K_{NASint}: key for NAS integrity protection in EPS, specific per selected integrity protection mechanism; shared between UE and MME;

K_{Upenc}: key for user plane encryption on LTE-Uu; shared between UE and eNodeB;

K_{RRCenc}: key for RRC message encryption, specific per selected encryption mechanism; shared between UE and eNodeB;

K_{RRCint}: key for RRC message integrity protection, specific per selected integrity protection mechanism; shared between UE and eNodeB.

4.8.3 Overall authentication and key agreement procedure in EPS (EPS-AKA)

This procedure is necessary at least when a UE enters initially an E-UTRAN domain (e.g. if the UE is switched on for the first time in this network), or any previously stored security context is no longer valid. A message exchange between UE and MME is used to create a fresh security context (see Figure 4-21). Before, it is necessary to provide an authentication vector to the MME from HSS.

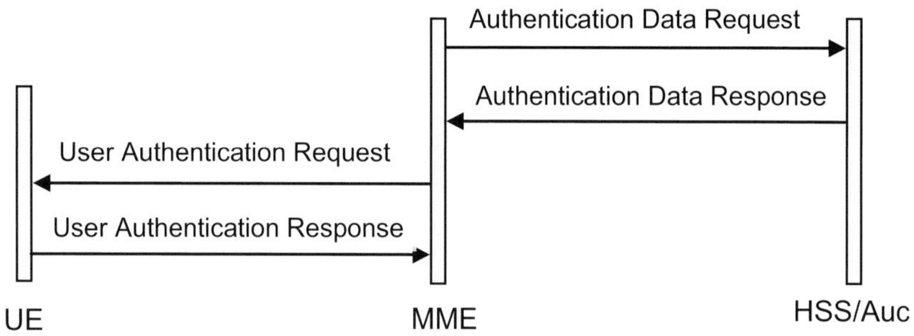

Figure 4-21: message exchange for authentication and key agreement in E-UTRAN

The result of an EPS-AKA and security mode command dialogue (see next sub-section) run is a valid EPS security context in UE and MME; it contains security keys and selected security algorithms. Temporarily, i.e. before the security mode command procedure has been run between UE and MME, a partial security context can exist. Such a context may also be mapped from a UMTS security context at intersystem handover. The EPS security context has an identifier assigned, by which it can be retrieved and used at a later point, without new authentication.

The contents of these messages are explained in the following table:

Authentication Data Request	IMSI: permanent UE identification. SN identity: identifies the serving network, consist of MCC and MCC Network Type: here set to E-UTRAN.
Authentication Data Response	Authentication vector(s): consists of the quadruple RAND, XRES, AUTN, K_{ASME} (see next two rows) and a numbering.
User Authentication Request	RAND: random number, used as input for the challenge to the UE AUTN: authentication token, KSI_{ASME}: a key set identifier;
User Authentication Response	RES: expected result of the authentication challenge

Table 4-5: contents of messages for download of authentication vectors and authentication and key agreement

4.8.4 NAS signaling security

The signaling between UE and the MME (NAS signaling, see explanation in sub-section 3.3.2 on the non-roaming architecture, and sub-section 6.9 on the protocol details) is critical from security point of view, because access control and resource allocation is provided through it, as well as private data which could be exploited or abused (e.g. service related information like APNs). It is therefore necessary to provide means to protect NAS signaling against eavesdropping and unwanted modification. Thus, ciphering and integrity protection is supported for the largeest part of NAS messages; the use of ciphering is optional and configurable by the operator by selecting the NULL ciphering algorithm. However, due to the need for initial handling (bootstrapping), at the point when a security context is not yet available between UE and MME, some messages may not be integrity protected.

As mentioned before, there is also some flexibility regarding the security algorithms, so that a specific NAS message pair is available for negotiating on them (supported and to be used); the message pair SECURITY MODE COMMAND / RESPONSE is counted as part of the EPS Mobility Management set of NAS messages.

4.8.5 Security for interworking with non-3GPP systems

4.8.5.1 General

A distinction has to be made, regarding the so-called "trust" property of the non-3GPP access network, via which the UE wants to access the EPC. The assignment of this property itself is up to the EPC operator (in case of roaming, to the home EPC operator), and there is no technical criterion defined for "trust" in 3GPP specifications. In fact, a non-3GPP access is identified as "trusted" simply if all the security features are deemed sufficiently strong (and how this is achieved is not standardized, rather it is a matter of inter-operator agreement); a non-3GPP access network may connect to different EPC's, and it might be deemed "trusted" by one and "untrusted" by a different EPC operator. The determination of "trust" is done either statically, by configuration data in the UE, or dynamically by an indication during access authentication (see sub-section 4.15.5).

The target security features to be provided are similar to those listed in sub-section 4.8.1; for user identity confidentiality again the concept of temporary identities is used (but identities themselves are different from those in the 3GPP access: EAP pseudonyms and fast re-authentication identities).

The security mechanisms described below are authentication and key agreement, and generate security contexts in the UE and in the non-3GPP access, which allow to provide user and signaling data confidentiality and integrity protection (but the detailed encryption and integrity protection algorithms are outside 3GPP's scope and thus left to the specification of the non-3GPP access technology). The encryption and integrity protection of signaling data between UE and the EPC is provided by the mechanisms described below, based on the security contexts in UE and EPC. Security between a node in the non-3GPP access (trusted or untrusted) and the EPC is implemented via the Network Domain Security concept (see [15]); it describes the use of Security Gateways and IPsec. Security within the non-3GPP access network is by nature out of scope here.

4.8.5.2 Security for interworking with Trusted non-3GPP networks

Authentication and key agreement is based on EAP-AKA prime ("Extensible Authentication Protocol – Authentication and Key Agreement prime") scheme, and therefore on EAP and EAP-AKA as the base protocols. EAP-AKA is widely used, e.g. also for 3GPP's I-WLAN (available since Rel. 6, offering already a "non-3GPP" access to 3GPP users, but without mobility). As such it was a natural choice to extend it generally for usage with non-3GPP accesses, and in a specifically modified version with trusted non-3GPP accesses.

The procedure is mainly run between the UE (in particular, the USIM) and the AAA server in the home EPC; in roaming a AAA proxy in the visited EPC is necessary (not shown). These two AAA nodes already exist for I-WLAN, and the procedures descriped below are only slightly extended. Between AAA server and the node in the trusted on-3GPP access (it acts as the EAP authenticator, see subsection 6.3) the EAP protocol messages are transported over (i.e. encapsulated in) DIAMETER AAA messages. HSS is involved at the backend for providing authentication vectors.

☞ See IETF RFC 3748 [16], RFC 4187 [17] and subsection 6.3 for more details on EAP and EAP-AKA; EAP-AKA prime is specified in IETF RFC 5448 [18].

The authentication and key agreement procedure, together with a re-authentication, for a UE in trusted non-3GPP access is shown in Figure 4-22 on a high level.

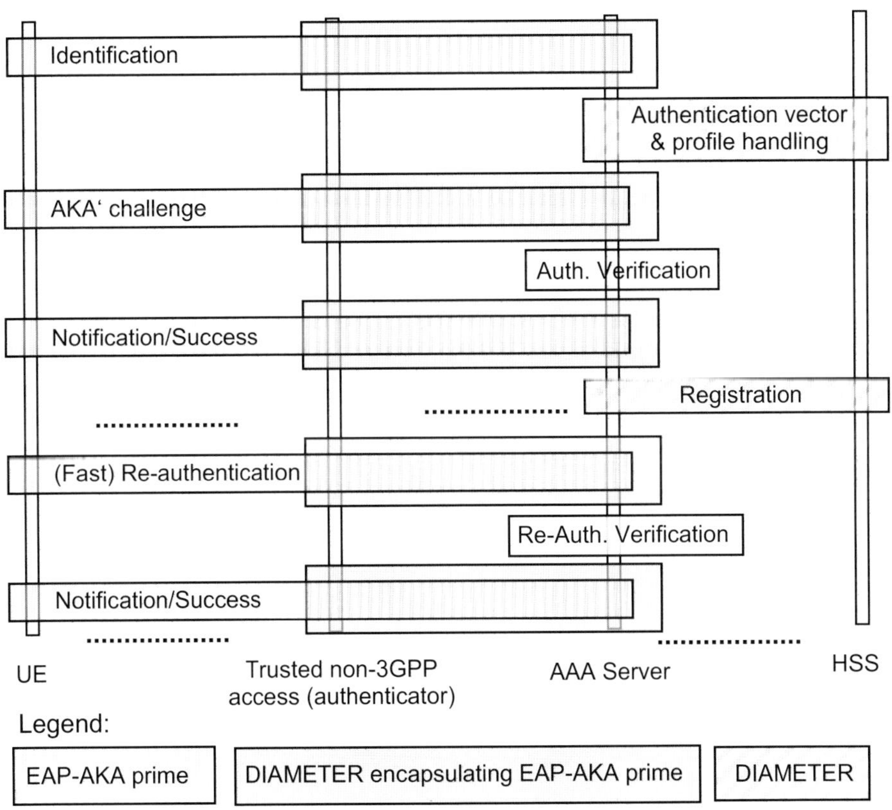

Figure 4-22: overall concept for authentication/key agreement and fast re-authentication in trusted non-3GPP access

The EAP-AKA prime procedure starts with identification of the UE; based on UE's identity the authentication vector and user profile data (i.e. subscription data) is fetched from HSS. Based on this data the UE can be challenged, and from its response verified that it is "authentic". The UE is then notified of its successful authentication and the registration with HSS can be performed. At some later point the UE can be asked for re-authentication (which can be "fast", if a fast-re-authentication identity had been assigned in the first authentication), and if successfully verified, this is notified again to the UE. The detailed procedures are presented in sub-section 5.1.2.

The re-authentication scheme mentioned here is called "fast", because it re-uses the keys derived in an earlier (full) authentication, and derives a new Master Session Key (MSK). It does not involve the original security credentials (i.e. USIM and authentication vectors), thus has fewer steps and messages to exchange. The fast re-authentication is mandatory for implementation in the AAA infrastructure, but optional to use (depending on the policy of the network operator; the use is indicated by providing a the fast re-authentication identity to the UE). The number of re-authentications is counted and limited (so that eventually a full authentication is triggered again, when the limit of fast re-authentications is reached).

4.8.5.3 Security for interworking with Untrusted non-3GPP networks

In the case of untrusted non-3GPP access the overall security procedure is conceptually similar, as far as the authentication and key agreement part is concerned; however, first of all it remains with the traditional EAP-AKA method (and **not** EAP-AKA', as for trusted non-3GPP accesses) and the difference in the interface and protocol used between the two networks (IPSec tunnel) is showing up. It is thus also termed "tunnel authentication and authorization". The overall procedure is shown in Figure 4-23.

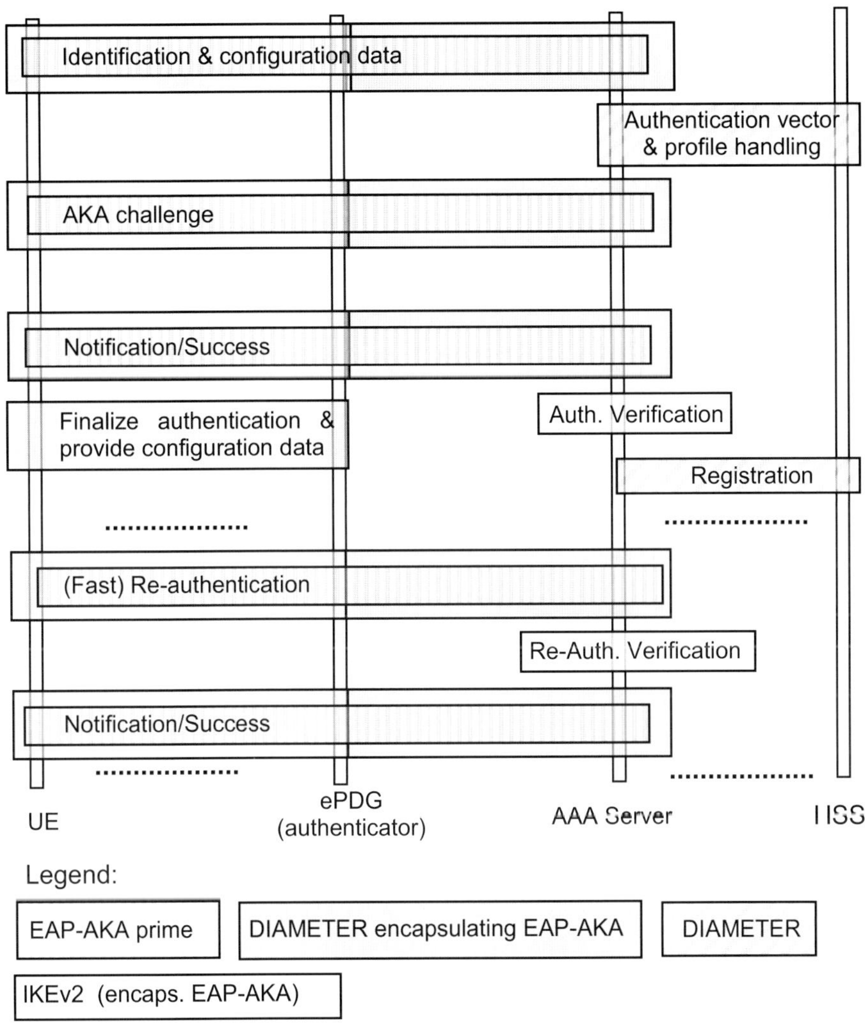

Figure 4-23: overall concept for tunnel authentication/key agreement and fast tunnel re-authentication in untrusted non-3GPP access

According to the architecture given in Figure 3-11 there is an IKEv2 signaling relationship between UE and the ePDG. It "emulates" a virtual layer 2 for the EAP, although it is actually on IP layer (= layer 3), by transporting EAP payload in IKEv2 configuration payloads.

4.9 EPS mobility and session management

4.9.1 General

A fundamental concept of the 3GPP system, valid from the first legacy 2G system to the SAE/LTE system, is the differentiation of idle and active mode. The intention is to optimize resources and terminal power consumption in idle mode, and it has a few drastic consequences:

1. the UE is in listening mode on the radio interface only and does not perform signaling with the network, as long as it remains in a defined area (the area is a list of assigned Tracking Areas in SAE/LTE, a Location Area in 2G and 3G CS, and a Routing Area in GPRS). If the UE moves out of this area, signaling for the sake of updating the network is performed.
2. the resources on the radio channels and between RAN and core network are deallocated. For originating communication the resources are allocated on request of the UE (service request), and the UE will then transition to the active mode.
3. for terminating communication the UE has to be paged first; thereafter a similar procedure as for originating traffic is executed to allocate resources.

In active mode the UE is able to transfer data immediately.

One should note that the concept of active and idle mode is assumed to be generally lacking in non-3GPP accesses. There are similar concepts in e.g. WIMAX and 3GPP2 systems; this fact is not exploited in the interworking scenarios with the former, but for optimized handover with HRPD the so-called "dormant" state is considered.

4.9.2 EPS mobility management (EMM)

Like in any control relationship, the UE and the network need to have a common and unambigous understanding about their states over time; a state model is defined for that purpose on NAS level, i.e. in the UE (see Figure 4-24, simplified) and in MME (see Figure 4-25). State transitions then occur as a consequence of NAS signaling between these two entities, or in some special cases also due to local decisions.

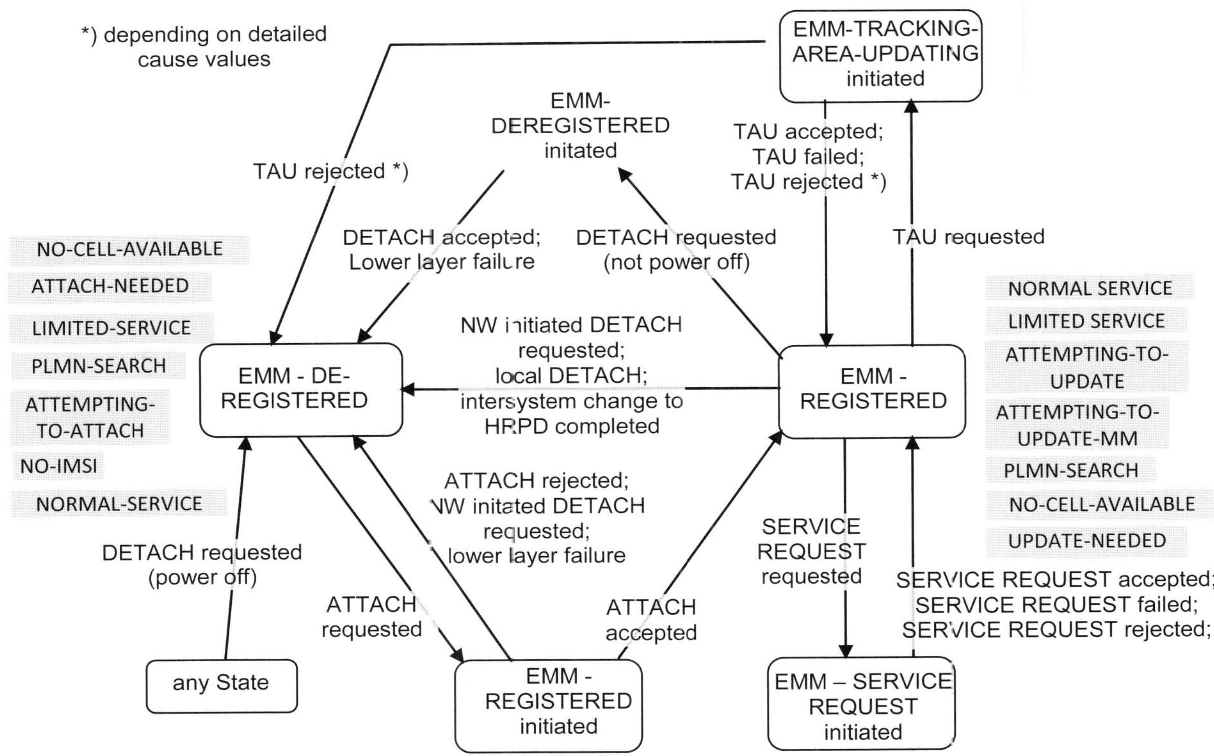

Figure 4-24: state model of EMM in UE

The two main states with a more extended duration are “EMM-REGISTERED” and “EMM-DEREGISTERED”; in the “EMM-REGISTERED” state the EPS has an established EMM context for the UE and knows the location of the UE either on the level of the assigned TAI list (in idle mode) or on the level of eNodeBs. In state “EMM-DEREGISTERED” the network has no EMM context, and thus the UE would first need to perform the attach procedure (which includes the registration with the network).

Four transient states exist, they are entered after NAS signaling requests have been made by the UE; depending on the outcome of the requests, the UE either falls back to the previous state (e.g. “EMM-DEREGISTERED”, if the attach procedure fails, or “EMM-REGISTERED” if the TAU was successful, simply failed, or was rejected only with uncritical cause), or transitions to the other main state as requested or needed (e.g. to “EMM-DEREGISTERED” if the TAU was rejected with a critical cause).

The UE states “EMM-REGISTERED” and “EMM-DEREGISTERED” are split up further into sub-states, as visualized in the graph.

Table 4-6 and Table 4-7 list further details about them, including some information on the selection rules.

Substate	Description
NORMAL-SERVICE	Entered as a consequence of selected negative outcomes of TAU and SERVICE REQUEST (e.g. implict detach by the network). The UE can try subsequently an attach.
LIMITED-SERVICE	Applies if the selected cell is in a forbidden PLMN or forbidden TA, or the associated CSG ID is not allowed for the UE. The UE needs to search for cells enabling normal service.
ATTEMPTING-TO-ATTACH	Applies if a previous attach has been rejected and the counter for this event is still below the limit.
PLMN-SEARCH	Selected initially after switch on, if the UE contains a valid USIM.
NO-IMSI	Selected initially after switch on, if the UE contains no USIM.
ATTACH-NEEDED	Can be entered for reasons like access class barring or rejection of NAS signaling connection establisment.
NO-CELL-AVAILABLE	Chosen f initial PLMN search was not successful; the UE continues the search for valid cells/PLMNs, but at a lower rate.

Table 4-6: UE’s sub-states in “EMM-DEREGISTERED” state

Note: in Rel. 9 some enhancements in the state descriptions can be expected for support of emergency services; e.g. a UE in “LIMITED-SERVICE” the UE will be able to attach for emergency services (but this feature will be supported only with Rel. 9).

Substate	Description
NORMAL-SERVICE	Initially chosen when the main state "EMM-REGISTERED" is entered, if no other reasons for other sub-states exist.The UE shall respond to paging and perform TAU, and it may send/receive data.
ATTEMPTING-TO-UPDATE	Entered if there was no response from the network to a TAU request. The UE shall perform further attempts for TAU, but cannot transfer any data.
ATTEMPTING-TO-UPDATE-MM	Entered if combined tracking area updating procedure was successful for EPS services only; this is related to coordination of mobility procedures between E-UTRAN and CS for the purpose of CS fallback (see e.g. sub-section 4.14.3). The UE can transfer user data and perform signalling.
LIMITED-SERVICE	See corresponding sub-state in Table 4-6.
PLMN-SEARCH	Aapplies during search for PLMN, due to the limit for unsuccessful TAU or SERVICE REQUEST attempts being exceeded.
UPDATE-NEEDED	Entered if the UE's access class is barred, or if the NE rejected the NAS signaling connection establishment; no user data transfer or signaling is allowed, except response to paging. The UE needs to perform cell re/selection.
NO-CELL-AVAILABLE	Entered if cell re/selection has not yet provided a suitable cell; the UE needs to perform cell re/selection.

Table 4-7: UE's substates in "EMM-REGISTERED" state

The EMM state model in MME is simpler; besides the likely longer term substates "EMM-DEREGISTERED" and "EMM-REGISTERED", equivalent to those defined on the UE side, it shows only two transitional states, one for the network initiated deregistration procedure, and one for all other initiated, common procedures.

☞ EMM procedures and related NAS signaling protocol details are specified in 3GPP TS 24.301 [19].

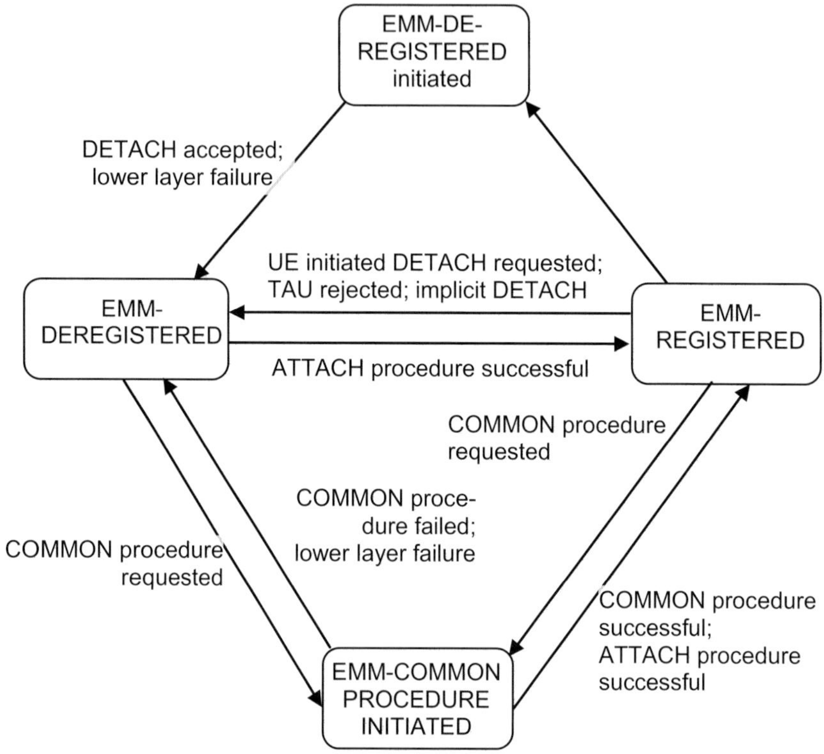

Figure 4-25: state model of EMM in MME

For the legacy PS system the equivalent state model is called GPRS Mobility Management (GMM). If the UE and the network support both of them, i.e. if interoperation with GPRS occurs, the question arises how these state models are interrelated. If the UE has performed successful attach in E-UTRAN, it enters the NO-CELL-AVAILABLE sub-state of GMM (it has the equivalent meaning as explained above for) and the NORMAL-SERVICE sub-state of EMM, and vice versa. If GMM is enabled in parallel to EMM, some coordination is also maintained with reject handling, e.g. with reject of an attach in E-UTRAN, the UE performs the handling of GMM state, GPRS temporary identifier (packet TMSI/P-TMSI), P-TMSI signature, RAI and GPRS ciphering key sequence number as it would handle them within GMM alone with the identical reject cause values.

Similarly, if CS Mobility Management (abbreviated simply MM) is also enabled, i.e. in a configuration with CS fallback, corresponding MM parameters like TMSI, LAI and ciphering key sequence number are handled in some reject cases. MM specific timers are generally not used while in E-UTRAN.

4.9.3 Idle mode process in UE

The UE has to perform the procedures of PLMN selection, cell re/selection, optionally CSG selection, and location registration, and the overall process scheme is seen from Figure 4-26 (a more detailed view is given in sub-section 4.19). This is largely the same as in the legacy 3GPP system, except for CSG handling being a new feature and E-UTRAN being another/new RAT.

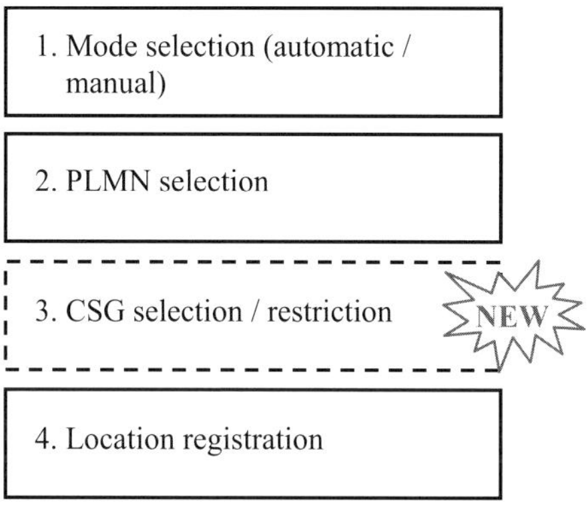

Figure 4-26: overall idle mode process in UE

For generality the term "location area registration" is used here, but it maps to the TAU procedure in case of E-UTRAN access.

There is a sharing of work between the radio access network (also called "Access Stratum", AS) and the core network ("Non-Access Stratum", NAS) in idle mode:

1. Actions on NAS level:
   - maintain lists for PLMN priority, associated RATs, forbidden PLMNs and equivalent PLMNs;
   - select a PLMN (and instruct AS to select a corresponding cell);
   - maintain lists of forbidden registration ares and allowed CSG ids;
   - select a CSG id (and instruct AS to select a corresponding cell);
   - perform NAS signaling for registration, TAU (i.e. change of registration area) and deregistration.
2. Actions on AS level:
   - search for (cells of) PLMNs, RATs and CSGs;
   - select and change (re-select) cells;
   - perform measurements and identify suitable cells;

- synchronization on radio channels of cells and monitor broadcast channels for information on PLMNs, CSGs;
- perform access class barring of UEs.

The last item (access class barring) is a feature present since Rel. 6 for legacy accesses. It can be applied in critical network conditions (e.g. disasters or high overload due to other networks' outages) and consists in a simple form in assignment of UEs to access classes (from 0 to 15) by static USIM configuration, and broadcasting of access barring indications for these access classes. Access Class barring is done only at initial connection requests, i.e. RRC CONNECTION requests, and it applies separately for the CS and (legacy) PS domain. Also the response to paging and location registration can be restricted by these access classes.

For E-UTRAN these main enhancements were done:

- a time duration (for which the access barring applies), and
- a percentage factor (of UEs for which the access barring applies)

may additionally be broadcast. Further extensions were under development at the time of writing for E-UTRAN in Rel. 9, due to the fact that in E-UTRAN all services are PS based and so a pure domain specific access control is no longer sufficient. The envisaged extension was that the barring can be applied selectively for Multimedia Telephony services (i.e. voice and/or video over IP under control of IMS).

☞ 3GPP TS 23.122 [20] describes UE's NAS procedures in idle mode; 3GPP TS 36.304 [21] contains the AS counterpart.The domain and service specific access control features are described in 3GPP TS 22.011 [22].

4.10 EPS session management (ESM)

Pushing the above explanations on EPS bearer further, there is a need for management of such bearers, i.e. their creation, deletion and modification. The underlying concept is called "EPS Session Management" (ESM). The corresponding overall state model is seen in Figure 4-27 (simplified; some error cases are left out).

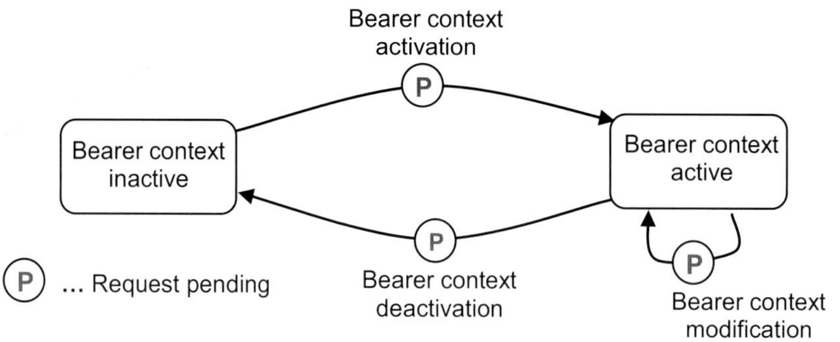

Figure 4-27: EPS session state model (in UE)

The (EPS) "bearer context" is the set of information describing the EPS bearer between the UE and the PDN; note that the EPS bearer context remains intact even if bearer portions (e.g. the radio bearer or S1 bearer) is temporarily not active. For the full bearer model two refinements are needed (but not elaborated here):

1. the bearer state changes may be triggered by the UE or by the network;
2. the procedures differ between default and dedicated bearers.

The UE may request allocation or teardown of resources (these resources being a PDN connections, or bearers). When a PDN connection is established also a default bearer is always created, and it exists until the PDN connection is released. Dedicated bearers are linked to the default bearer of the same PDN connection. Both default and dedicated bearers can be modified; however, the UE can only ask for the amount of resources, it is the networks decision whether an existing bearer is modified or a new one is established. An overview of the full set of ESM related messages is given in sub-section 6.9.3. IP address allocation, an integral part of PDN connection establishment, is presented in sub-section 5.3 (due to its linkage with procedural aspects).

Here we explain one specific ESM procedure in more detail, namely the UE requested PDN connectivity; it may be piggybacked onto the EMM attach procedure (then the resulting PDN connection would be the first one for the UE, and no APN is needed, i.e. default PDN connectivity applies), or standalone (in case of multiple PDN connectivity, and an APN is required to specify the desired PDN). The structure of the message flow is:

UE → PDN CONNECTIVITY REQUEST → MME

UE ← ACTIVATE DEFAULT EPS CONTEXT REQUEST ← MME

This procedure is supervised by a timer, so that after timer expiry without the expected response the error handling is entered. The PDN connectivity request bears an indication whether it is for the first time (while attached to EPS), or a subsequent one (as it occurs in course of handovers back to E-UTRAN access, after the UE was intermediately in a non-3GPP access (but still under the umbrella of the EPS!).

Within this transaction the UE may also request to receive so-called "Protocol Configuration Options" (PCOs) from the MME. The concept of PCOs has been carried over from the legacy GPRS, where a frequent usage was to request a P-CSCF or DNS server address, or to indicate IMS signaling for the PDN connection. With EPS it was extended for DSMIPv6 Home Agent address and Home network prefix delivery, as well as indication whether immediate IPv4 address allocation (via NAS signaling) or defered (via DHCPv4), see also sub-section 5.3.2. PCO transfer (request by UE and response by the network) is possible in several other ESM message dialogues (EPS bearer context de/activation, EPS bearer context modification, PDN connectivity/disconnect request, bearer resource allocation/modification).

If the PDN connectivity is granted only with a restriction in IP version (e.g. if a dual stack PDN connection was requested but the network policy allows only one particular or only a single IP version), this is indicated by a corresponding cause value in MME's response message ACTIVATE DEFAULT EPS CONTEXT REQUEST.

☞ More details on the ESM model and related NAS signaling protocol details are found in 3GPP TS 24.301 [19].

4.11 Quality of Service (QoS) Concept

4.11.1 General on QoS in the Evolved 3GPP System

3GPP's concept of Quality of Service in the legacy PS domain is built on four traffic classes (conversational, streaming, interactive and background) and offers around a dozen independent parameters for PS bearers (which are set up in the form of PDP contexts). But some of these QoS parameters became obsolete; over the years also advances in transmission/transport technology were achieved, and vast experience from deployments with mass usage became available. As a result, a strong demand emerged to simplify the QoS concept for SAE.

☞ The legacy QoS framework is specified in 3GPP TS 23.107 [23].

The Evolved 3GPP system supports UE initiated and network initiated QoS control. Overall, the support and utilization of differentiated QoS is expected to show a time dependence roughly as shown in Figure 4-28 (arbitrary units are used for both axes): QoS in general is taken on first in a UE controlled manner, and only gradually. From a certain time onwards network initiated QoS control is started to be introduced and spreads in the networks. This will lead to a saturation or even decrease of the UE controlled QoS mode.

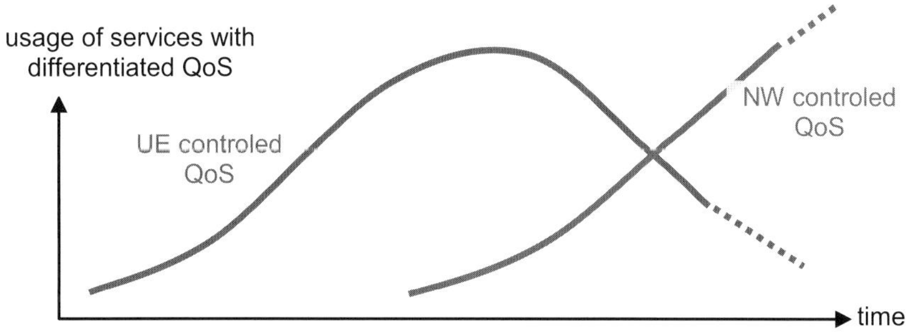

Figure 4-28: expected evolution of QoS control mode usage over time

For the sake of differentiated and flexible operator control, the QoS concept needs to be applied both on the service and the access level. E.g. for VoIP service some QoS with real time characteristics is required end-to-end, which is to be negotiated and/or assigned. An access network with QoS support (e.g. if it is involved on the originating side of the communication) then needs to be informed about this decision, so that it can map the end-to-end QoS parameters onto access specific bearer handling. In the Evolved 3GPP system two methods are used for such signaling:

1. on path: this is together with signaling for installing the traffic path, which requires mobility related signaling (realized in the case of GTP as the EPC protocol for S5/S8).
2. off path: independent of the traffic path; this variant applies with MIP (all variants, as they do not include QoS support).

QoS can be provided in a static or dynamic way; static QoS would be stored as subscription data in HSS and e.g. distributed as a profile data via the AAA signaling (see sub-section 4.7). But for the Evolved 3GPP system there is also a strong need to provide QoS dynamically, i.e. dependent on the ongoing service. Dynamic QoS, as the special added value provided by the mobile network and service operator, is then always assumed to be coupled with differentiated charging. Both together are handled by dynamic policies, see sub-section 4.12.

4.11.2 QoS Parameters in EPS

LTE/SAE utilizes a class based QoS concept, which reduces complexity while still allowing enough differentiation of traffic handling and charging by operators. The QoS profile for an individual EPS bearer includes the following parameters:

1. QCI (QoS Class Index): the interpretation is not standardized, but left to the network(s); in roaming, where potentially two EPSs are involved, it will depend on roaming agreements. Typically QCI will be used by eNodeBs for IP packet forwarding over the radio link (e.g. scheduling) and potentially within the EPC nodes for rate control.
2. ARP (Allocation Retention Priority): it contains the priority level (a number) and the pre-emption capability and vulnerability (two flags). It is relevant in situations of scarce resources, where the priority level is used to decide if an additional bearer can be established or an existing one modified to higher resource level, as well as if a bearer can be dropped in exceptional cases like handover. The pre-emption capability flag determines whether another bearer can be dropped due to the needs of the considered one, the pre-emption vulnerability flag encodes the opposite (whether the considered bearer can be dropped due to needs of other bearers). ARP is not sent to the UE, as it is only used within the network.

GBR-bearers (those with guaranteed resources) are additionally characterized by two parameters:

1. GBR (Guaranteed Bit Rate): it is the steadily maintainable bit rate for a GBR-bearer. GBR cannot be adjusted by the radio access (if radio conditions do not allow to maintain the GBR, it will be released).
2. MBR (Maximum Bit Rate): it determines the upper limit of traffic provided through a GBR bearer; the effective use of this parameter can be foreseen for the future, currently the MBR is set equal to GBR. For traffic flows exceeding the MBR, packets are subject to drop.

Also two parameters are assigned with the sets (or "aggregate") of EPS bearers of type "non-GBR":

1. APN-AMBR: aggregate maximum bit rate for all non-GBR EPS bearers for a PDN connection, as identified by the APN and stored as subscription parameter in HSS. It is enforced by the PDN GW in uplink and downlink, and by the UE in uplink.
2. UE-AMBR: aggregate maximum bit rate for all non-GBR EPS bearers of a UE, defined by subscription in HSS. It is enforced in up- and downlink in the eNodeB.

Each of the non-GBR bearers could potentially consume the whole APN-AMBR, or even UE-AMBR (e.g. if no other non-GBR bearers are at this moment used for data transmission). Excess traffic may be discarded by shaping/policing function in the involved mentioned user plane nodes (UE, eNodeB and PDN GW). All the bit rates in QoS parameters refer to the traffic on the S1-U reference point without GTP/IP overhead.

4.11.3 Mapping between pre-Rel. 8 QoS and QCI

When an intersystem change occurs, it is necessary to map between legacy and new QoS parameter sets, see Table 4-8. Pre-emption capability and the pre-emption vulnerability are discarded in mapping from EPS to pre-Rel. 8 QoS. In the direction from pre-Rel. 8 to EPS their settings are not standardized and left to the operator's configuration.

QCI	Traffic Class	Traffic Handling Priority	Signaling Indication	Source Statistics Descriptor
1	Conversational	not applicable	not applicable	speech
2	Conversational	not applicable	not applicable	unknown
3	Conversational	not applicable	not applicable	unknown
4	Streaming	not applicable	not applicable	unknown
5	Streaming	1	yes	not applicable
6	Streaming	1	no	not applicable
7	Streaming	2	no	not applicable
8	Interactive	3	no	not applicable
9	Background	not applicable	not applicable	not applicable

Table 4-8: Mapping between standardized QCIs and pre-Rel-8 QoS parameter values

Between EPS ARP priority and pre-Rel. 8 QoS parameter "ARP" the mapping is given by Table 4-9; H ("high"), M ("medium") and L ("low") are priority levels for operator's choice, obeying the relation 1 <= < H < M <= L <=15.

Pre-Rel. 8 ARP	EPS ARP priority	EPS ARP priority	Pre-Rel. 8 ARP
1	1	1 to H	1
2	H+1	H+1 to M	2
3	M+1	M+1 to 15	3

Table 4-9: mapping between EPS and pre-Rel-8 ARP values

The detailed rules (including all aspects of SGSN and PDN GW handling) are quite complex and not given here (see 3GPP TS 23.401 [24]). Mapping is done one-to-one between an EPS bearer and a PDP context, i.e. there is no splitting of bearers.

4.12 Concept of Policy and Charging Control (PCC)

4.12.1 Evolution up to Release 7

In 3GPP Rel. 6 two parallel concepts were present: Service Based Local Policy (SBLP) and Flow Based Charging (FBC, defined in [25]), see Figure 4-29. It was then found that these two concepts have much in common, and they were unified in Rel. 7 in a the Policy and Charging Coordination (PCC) framework; Rel. 8 evolves it further and adds several enhancements.

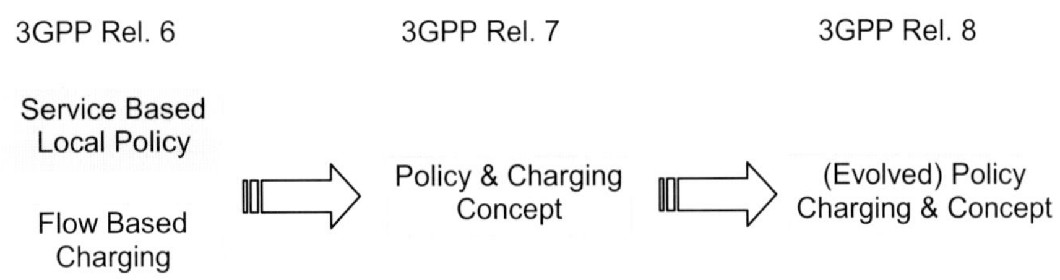

Figure 4-29: Evolution of policy and charging related concepts in 3GPP

The term "policy" is used very generally, already in 2G/3G technology (see [26] in Rel. 7 for background); two sub-variants are

- operator's policy: there is a wide range of options how an operator may want to differentiate users and their data traffic;
- service based policy: the type of an invoked service determines how user data transmission occurs.

Two types of policies are relevant:

- QoS policy: it describes how QoS parameters shall apply to a packet flow, and
- "gating" or "filtering" policy: defines whether a packet flow is allowed or not.

The subject of policies are are so-called Service Data Flows (SDFs), defined by sets of IP packet filters; these in turn are described by patterns of IP header fields and, to some extent, payload. To this end, the PCC provides means for deep packet inspection in the user plane. Additionally, the functionality of event reporting from SDFs is included.

The motivation for unification of these policy aspects with charging, as done in Rel. 7, was that they have components in common: filters, rules and events. The overall context for these is shown in Figure 4-30 for the base scenario. The Policy and Charging Rule Function (PCRF) contains all the intelligence needed to decide on what kind of IP traffic is allowed for a user, with which kind of charging and under which conditions of service delivery; it is a pure control plane node, with a rule DB and potentially interfacing with a subscriber repository and Application Functions (AFs; in most cases, but not necessarily, these are IMS applications, with Proxy-CSCF being the AF representative). It can instruct the Policy and Charging Enforcement Function (PCEF) in the user plane dynamically with the rules. The PCRF concept was formulated generically, for any IP Connectivity network (IP-CAN).

Example locations of PCEF in Rel. 7 are:

- GGSN (for GPRS as IP-CAN);
- PDG (for I-WLAN as IP-CAN), in the Cable Modem Termination System (for the packet cable technology DOCSIS);
- a Gx-capable node in the WIMAX ASN or WIMAX CSN.

The node hosting the PCEF also provides the interfaces with Online Charging system (OCS) and Offline Charging system (OFCS). Note that more than one PCEF and/or PCRF could be involved for one user. Multiple PCEFs come into play e.g. if several APNs are used and they are hosted on different GGSNs; multiple PCRFs may be employed for reasons of load sharing in a larger network.

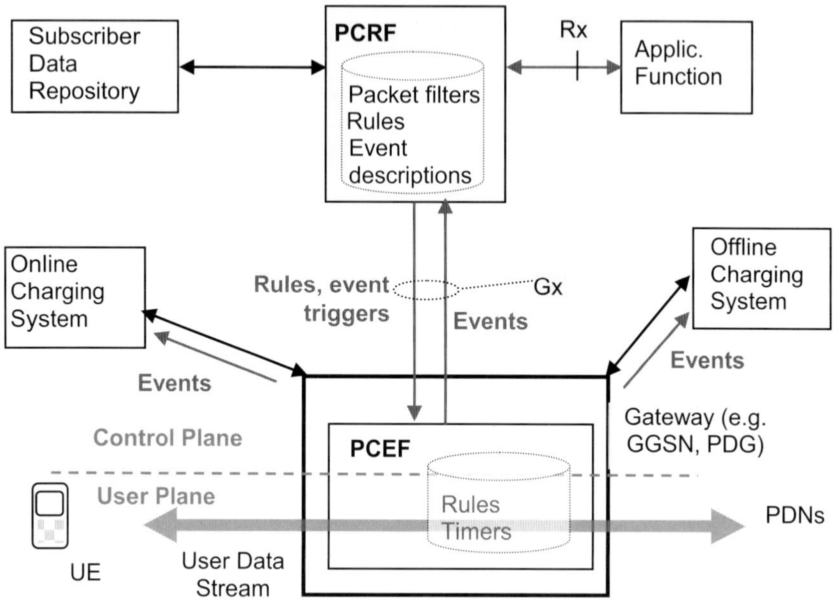

Figure 4-30: legacy (Rel. 7) concept for Policy and Charging Control

For the intended working of the PCC it is necessary to bind the AF session to the IP-CAN session, which is a 1:1 mapping; UE's identity, IP address and used APN are considered herein. However, an AF is not always involved (as it is the case with pure IP traffic), and PCRF may then act based on its own data. Next, the PCC rules are generated, QoS parameters are assigned by the PCRF and transferred to PCEF. Finally the PCC rules need to be bound to bearers; this could involve the modification of existing bearers or the creation of new ones. Additionally, usage reporting (for the traffic passing through, in a differentiated manner) to online and offline charging systems, together with event reporting (also to PCRF) is performed by the PCEF; for that purpose the PCRF "subscribes" to the events happening in the user plane, or simply requests time based re-validation of rules.

4.12.2 PCC in Release 8

The evolved (i.e., Rel. 8) PCC builds upon and extends this concept; EPC is seen as a new IP-CAN, with PCEF located in PDN GW.

An enhancement for Rel. 8 was necessary due to the need for support of IETF based EPC protocol architecture variant (i.e. for PMIP based S5/S8). In this case the bearer binding occurs for the GTP access on the Serving GW (in contrast to the GTP based architecture variant, where this happens on the PDN GW). The logical function "Bearer Binding and Event Reporting Function" (BBERF) was introduced for this purpose, located on the Serving GW (see Figure 4-31, where the unnecessary details like subscriber data repository and charging functions are left out). Its service data flow detection functions correspond to those of the PCEF for GTP, with the extension that tunneled data flows can be handled as well. Additionally, BBERF reports events to the PCRF and bearer bindings are also verified in uplink direction. For PCRF with trusted non-3GPP access, BBERF is also located in the gateway node providing the MAG function. The control relationship between BBERF and PCRF is called GW control session.

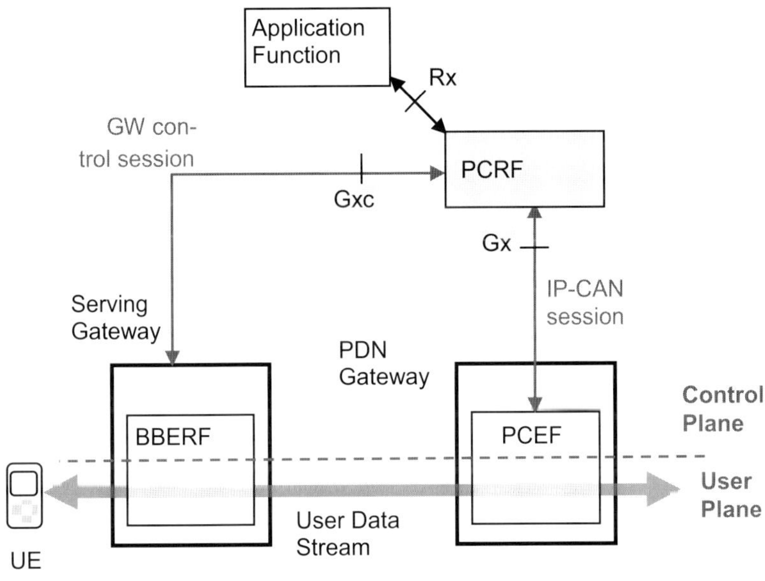

Figure 4-31: evolved concept for Policy and Charging Control (PMIP as EPC mobility protocol)

The case of DSMIPv6 requires yet another treatment, e.g. if mobility via S2c is used in trusted non-3GPP access. Then the Care-of-Address will be used in the IP header of the DSMIPv6 tunnel, and this must be known at the BBERF. In conclusion, these three different cases need to be considered:

1. No GW Control Session is required: this is 3GPP access where GTP-based S5/S8 are employed.

2. A GW Control Session is required. The BBERF establishes it prior to any IP CAN session establishment. There are two sub-cases:

   a) The UE acquires a care of address (CoA) that is used for the S2c reference point. The same GW Control session applies for all IP CAN sessions using that CoA; BBERF does not provide an APN.

   b) Each IP CAN session is handled in a separate GW Control Session; BBERF provides an APN. PCRF expects tunnelling header information for each IP CAN session to be provided by the applicable PCEF.

Another extension is required for roaming with local breakout, in which case two PCRFs may be involved, optionally one in HPLMN and one in VPLMN (see Figure 4-32, which also shows a BBERF in VPLMN, indicating that it is suitable for a PMIP based EPC). PCC rules are exchanged between PLMNs, and policies from both HPLMN and VPLMN can be combined. For clarity the charging functions are shown again: the OCS is still located in the HPLMN with a roaming interface for real time control. OFCS now resides in the VPLMN and the collected charging data is exchanged in non-real time (in bulk mode) via specialized interfaces. In a simplified variant, the services could also be delivered under purely VPLMN based PCC.

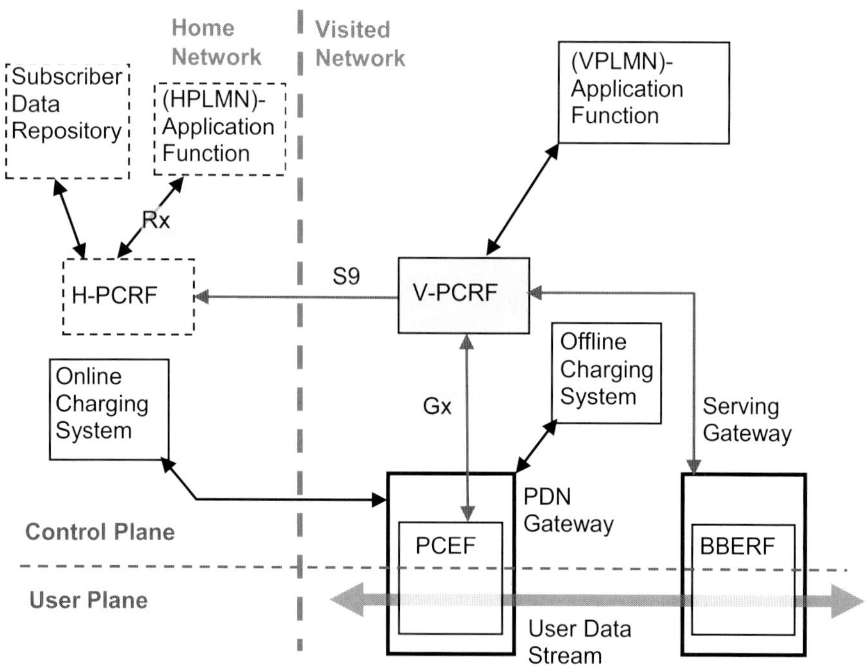

Figure 4-32: evolved architecture for Policy and Charging Control (roaming with local breakout)

In Rel. 8 also CDMA 2000 © HRPD was defined as another IP-CAN for PCRF, and at the time of writing efforts were on the way to align or interwork 3GPP's PCRF concept with the policy management scheme of TISPAN.

☞ The PCC framework is specified in 3GPP TS 23.203 [26].

4.12.3 Packet filters, PCC rules and events

Packet filters may consist of different patterns, as listed in Table 4-10:

Pattern Nr.	Constituents
1	quintuple of source IP@, destination IP@, source port number, destination port number, protocol ID (of the protocol "above" IP, i.e. of the next level payload); extension with Type of Service/Traffic Class *)
2	destination IP@, protocol ID of the protocol ID, Type of Service/Traffic Class *), IPSec Security Parameter Index (SPI);
3	destination IP@, the Type of Service /Traffic class *), Flow Label (in IPv6)
4	extended filters for transport and application layer (for deeper packet inspection, e.g. for for HTTP, WAP); these filters must be predefined in PCEF.

*) depending whether IP@ version IPv4 or IPv6 is used

Table 4-10: patterns for packet filters

Packet filters may include also ranges (e.g for port numbers) and masks (for IP addresses and Type of Service/Traffic Class parameters. A packet filter may also include a wildcard, which makes it a "match all" filter. A complete Service Data Flow is then defined by a set of packet filters (described by a SDF Template), with packet filters for downlink and uplink;

Events are typically changes in the PLMN, location (on different levels of granularity), QoS or IP address.

A PCC rule consists of a name, a service identifier, SDF filter, precedence of the rule (in relation to others), gate status ("open" or closed"), QoS parameters, charging key (mapping to e.g. a tariff). Three operations are defined for the signaling between PCRF and PCEF/BBERF: installation, modification, removal). PCC rules may be statically provisioned (in the PCEF), or dynamically provided (signaled from PCRF).

An example of PCC usage would be: the network operator does not charge for IP traffic of "control" type, e.g. DNS and DHCP traffic within the own network domain. The relevant port numbers 53 (DNS) and 67/68 (for DHCP request and response) as well as the subnet mask for the operator's IP domain (where the DNS/DHCP servers will be located) are included in a packet filter; this packet filter is then included in a rule including a charging key resulting in zero charge.

4.13 Idle Mode Signaling Reduction

Idle mode signaling reduction (ISR) is a mechanism to optimize, i.e. reduce, the signaling for both the UE and PS domain nodes when the UE is in idle mode; it is applicable if both of the two 3GPP PS accesses (E-UTRAN and GERAN / UTRAN) are available. In this case it should not be necessary to involve the core network if the UE alternately selects a cell of one or the other and moves along in idle mode. At this point it is worthwhile to recapitulate the idle mode behavior of a UE in a 3GPP system.

In one domain, CS (GSM or UMTS) or PS (GPRS or UMTS), the UE runs a quite elaborate process to find always a suitable cell; main criteria are the PLMN of a cell (obviously the HPLMN or an equivalent one will be preferred whenever available) and the radio signal strength, but cells may also be restricted for access, and cell re-selection should not be too often. If a cell is selected, the UE tunes to the broadcast, paging and common control channels (this is termed "camping"). It need not notify the core network as long as the new cell is in the same Location Area (LA, for CS access) or Routing Area (RA, in PS access); in the other case a Location Update or Routing Area Update (signaling on NAS level between UE and the core network) must be done. In this way the core network knows the location of the UE on the granularity of a LA or RA. As shown in sub-section 4.5 on the tracking concept, the concept was extended, with some differences, to E-UTRAN.

For a intersystem change in idle mode from/to PS GERAN/UTRAN access to/from E-UTRAN access the UE normally would have to register with the system it intends to use; this is necessary mainly to ensure that the paging is sent to the system where the UE is camping on. But with ISR, the UE is registered to both systems, legacy PS and the EPS, and paging will be distributed to both. Still then there are special situations, where - due to a need to synchronize all three entities, UE, MME and SGSN - ISR needs to be deactivated, so that signalling is again triggered with the appropriate node in the other core network. The UE runs two parallel timers for the update signaling/registration.

Figure 4-33 and 4-34 compare the two idle mode mechanisms in EPS, with and without ISR.

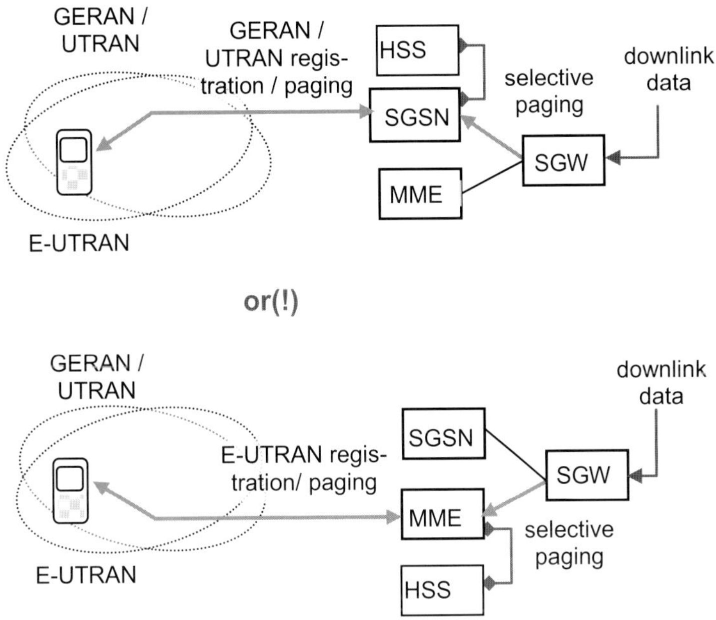

Figure 4-33: idle mode without ISR

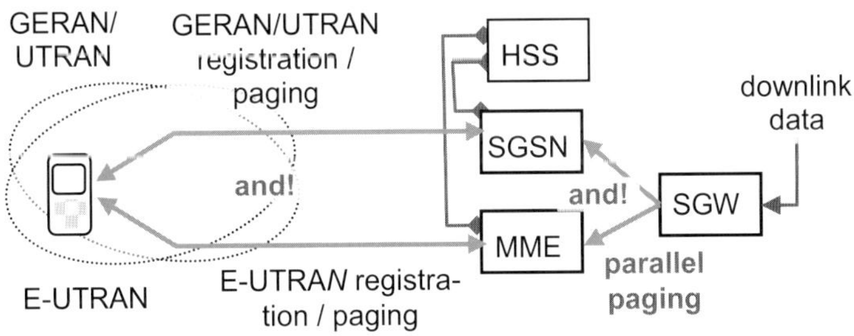

Figure 4-34: idle mode with ISR

The core network decides on the overall level about the usage of ISR, but the UE may prefer, due to a need for special handling, the ISR usage. The UE holds a parameter (TIN – "temporary identifier used in next update") indicating which type of mobility management context shall be used in the next signalling with the core network, which can be RAU or TAU. Possible values of TIN are "GUTI" (UE's identification in E-UTRAN/EPS,see 0), "P-TMSI" or "RAT related TMSI"; from that the activation status of ISR can be deduced, see Table 4-11. Only if for the next pdate signaling TIN is set to "RAT related TMSI", ISR is activated in the UE.

ISR activation in (successful) CN to UE signaling	**Signalled in RAU or TAU?**	**Current TIN setting in UE**	**TIN setting for next RAU/TAU exchange (UE to CN)**	**Meaning**
no	TAU	any	GUTI	ISR deactivated
no	RAU	any	P-TMSI	ISR deactivated
yes	RAU	GUTI	GUTI	ISR deactivated (UE has locally suppressed ISR)
		P-TMSI or RAT related TMSI	RAT related TMSI	ISR activated
yes	TAU	P-TMSI	P-TMSI	ISR deactivated (UE has locally suppressed ISR)
		GUTI or RAT related TMSI	RAT related TMSI	ISR activated

Table 4-11: TIN (meaning and setting in UE)

ISR as a feature is mandatory for E-UTRAN UEs supporting also GERAN or UTRAN (or both), but optional for the core network (on an overall or per UE basis).

Figure 4-35 explains the procedure for ISR activation, performed between UE, MME, SGSN and HSS.

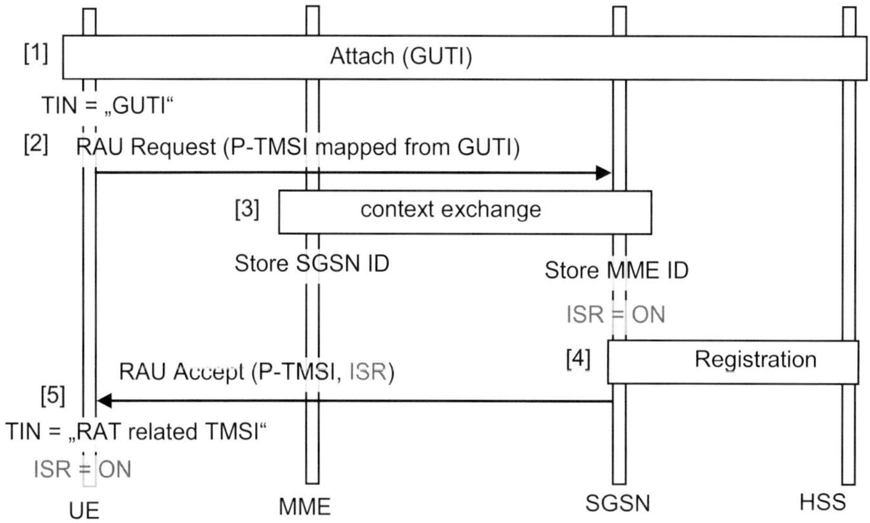

Figure 4-35: ISR activation

The necessary steps are:

1. Normal attach in E-UTRAN; any ISR setting is cleared, else no ISR related action (as ISR is never activated after within the attach procedure). As a result, the UE sets its TIN to "GUTI".
2. The UE chooses to select now GERAN/UTRAN as its desired access (althoug it stays in idle mode); it sends a RAU request to SGSN, containing P-TMSI (which is mapped from the GUTI).
3. The SGSN fetches from MME UE's context; MME indicates ISR support. The two nodes store their relationship. At this point SGS also establishes the necessary, parallel control plane connection with the Serving GW (not shown here). SGSN marks the ISR activation.
4. Registration of SGSN with HSS is done.
5. In the RAU accect message from SGSN also an indication for ISR activation is included. Based on that, the UE sets its TIN to "RAT related TMSI" and as a consequence ISR is now activated.

4.14 Interoperation with legacy circuit switched domain

4.14.1 General

With the advent of EPS – as a purely packet switched technology – the strong need arises to interoperate with the huge installed base of CS based networks. The most successful service, and therefore the one generating the largest share of revenue for mobile operators, is still the voice service (telephony); and in 3GPP's 2G and 3G technologies it is realized via CS. And for some time it will still be the legacy CS infrastructure which provides the overall and widest radio coverage. This investment of operators is to be protected.

Specialized solutions for interoperation with legacy voice services, for different deployment and usage scenarios, were proposed and finally standardized in 3GPP Rel. 8:

- For the case of localized E-UTRAN deployments within extended 2G/3G CS coverage, and mobility across the boundary of E-UTRAN coverage during a voice call: the assumption is that (1) voice over IP (via IMS) is deployed by the operator, and (2) due to the limitations of radio bands, no parallel operation of radio transceivers (for both radio technologies) is possible; the solution is called Single Radio Voice Call Continuity and is based on an application server in the IMS domain, from where interworking towards MSC is controlled.
- For the case of overlapping coverages of E-UTRAN and 2G/3G CS, where the operator does not (yet) want to deploy IMS for voice services, and the decision on the voice call handling is taken before the corresponding originating or terminating request is made/accepted: a fallback mechanism to CS has been developed (CSFB/CS fallback). Services related to voice (SMS and Supplementary Services) were included in this mechanism.

Both solutions have variants for the legacy CS technology stemming from 3GPP or from 3GPP2 (i.e., 1xRTT). Altogether they constitute high investments into standardization, development, testing, rollout and deployment/operation; however, one can see from this fact also their importance for operators.

Another solution for voice in LTE/EPS was proposed [27], but finally not followed on in 3GPP: CS over PS, meaning that the legacy CS protocol stack is left unchanged, and the CS signaling is only encapsulated and transported over the Evolved PS layer to a the point of interconnection (enhanced MSC), where it is then de-capsulated and handed over to the conventional CS protocol machine again.

4.14.2 SRVCC (Single Radio Voice Call Continuity)

As explained above, a concept for smooth interworking at least for ongoing voice services between the PS system and legacy CS system (both in its 3GPP and CDMA2000 © instantiation, 1xRTT) was sought. It relies on standard means of call control in the PS domain, which is realized by 3GPP's "Internet and Multimedia Subsystem" (IMS). This is termed "Voice Call Continuity" (VCC), and there are two variants:

1. Based on dual radio: it can be applied when the characteristics of the two involved radio technologies allows parallel operation of transceivers (senders and receivers). A solution was developed already in Rel. 7 for VCC between legacy CS and pre-SAE PS technology (in practice WLAN). In such a case the two systems need not be coupled tightly, because handover preparation for the target system can take place while the terminal is still connected via the source system. Dual radio VCC is not discussed further here.
2. For single radio (SRVCC): if the bands of the two involved radio technologies are too near, due to physical limitations of radio transmission (interference), transceivers cannot operate in parallel. In this case the handover must be prepared in the target access while the terminal is connected to the source system, because no connection can be opened in parallel. It leads to a tighter coupling than variant 1, both in terms of interfaces and procedures.

Based on the expected rollout scenarios of E-UTRAN, only the direction from E-UTRAN PS → CS was considered for SRVCC. The underlying assumption is that CS coverage exists virtually everywhere, and then there is no need to continue a CS call in PS, if the corresponding coverage is entered. The case of E-UTRAN only coverage is thus not covered. The basic concept is depicted in Figure 4-36 for CS being a legacy 3GPP system.

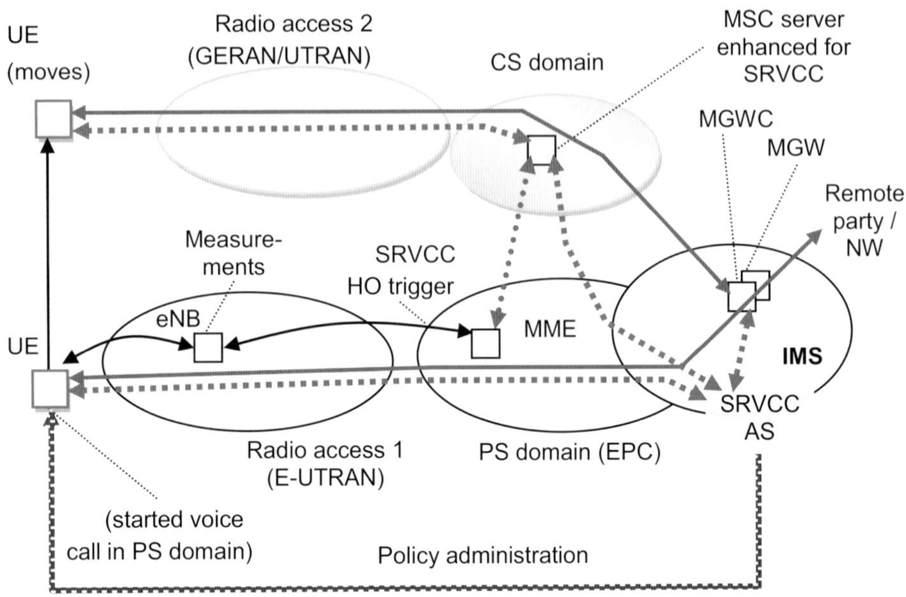

Figure 4-36: basic scenario for SRVCC (E-UTRAN to GERAN/UTRAN)

The concept requires:

- a specific enhancement on the CS control node in the legacy CS domain (MSC in 3GPP system, 1xIWS [Interworking Solution function] in the CDMA 2000 © 1xRTT case);
- an application server (AS) for SRVCC; and
- enhancements in the UE.

The building blocks of the solution are depicted on a high level in Figure 4-37. The voice call is anchored in IMS before SRVCC handover; as the UE moves, it performs measurements of the radio signal (strength and quality of transmission), both for the E-UTRAN cell it is currently using and the neighbouring GERAN/UTRAN cells (it knows these thanks to instructions by the serving cell). The measurement reports are sent to the serving cell and used to determine the handover point, and eventually the concrete cell to be used after handover; in this process also indications about the ongoing voice call, UE's SRVCC support and voice support of GERAN/UTRAN cells are utilized. The eNodeB then sends a SRVCC trigger message to MME. MME initiates the intersystem handover via a trigger message across Sv, which leads to resource reservation in GERAN/UTRAN like for an intra-CS handover. Also, the MSC server initiates signaling for call establishment towards the SRVCC AS. After resource reservation has been acknowledged back to MME, it commands the UE to move to GERAN/UTRAN. The UE performs there CS handover completion. The full message flow details is presented in sub-section 5.13.

The SRVCC solution does not only support a single voice session to be handed over seamlessly. It can also be used if the voice call occurs in combination with a non-voice service, like video streaming, although in this case the non-voice traffic handover does not enjoy the same smoothness/seamlessness. Still, the user experience for voice is most critical, and small interruptions of e.g. the video or fallback to lower bandwidth are deemed acceptable. With non-voice components, a PS handover procedure is performed in parallel to the PS-CS handover.

The policy of the network operator with respect to UE's SRVCC behaviour is administered by OMA Device Management and corresponding data is transferred to the UE; a specific Management Object has been defined, containing settings for numbers to be used in session transfer requests, preference of transfer directions, preference of media in different access systems, detailed conditions on when the transfer shall take place.

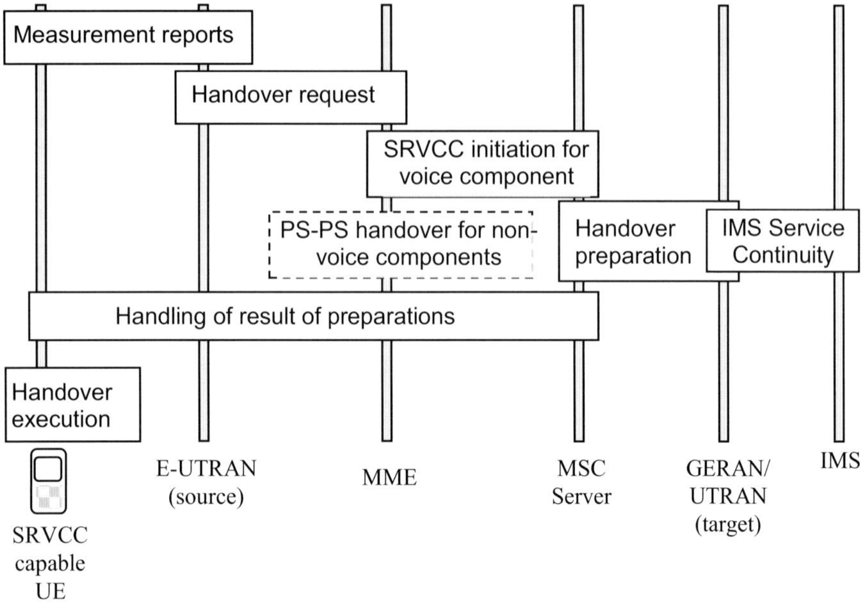

Figure 4-37: building blocks for SRVCC (from E-UTRAN to GERAN/UTRAN)

☞ See 3GPP TS 23.216 [28] for the functional specification of SRVCC.

4.14.3 CS Fallback from E-UTRAN to GERAN/UTRAN

Another solution for the case where IMS is not deployed, or not preferred, and with some limitation (namely, that the voice communication is not yet established) is possible by CS Fallback (CSFB).The mechanism is explained here for fallback from E-UTRAN to legacy 3GPP access (i.e. GERAN/UTRAN), but it applies to fallback to CDMA2000 © 1xRTT access similarly.

EPS, as a pure packet switched technology, would in principel foresee IMS (i.e. the packet switched control/signaling layer) for handling sessions. But during the design phase of the Evolved 3GPP system operators differed in their envisaged deployment schemes: some of them were hesitant to couple IMS deployment with LTE radio access and EPC core deployment. Instead, a smooth transition from already deployed CS infrastructure was sought. As a consequence requirements were formulated to have the possibility to fall back to CS under the following boundary conditions:

- the UE is under combined coverage of either GERAN or UTRAN and E-UTRAN,
- the feature is applicable for originated or terminated CS voice calls and SMS, and network triggered SS handling.

Even if IMS is deployed, it could also depend on preferences of the operator which kind of terminals should utilize CS or PS control for CS type of voice and related services. The fallback mechanism is provided for the case that the UE is registered in LTE/EPC and one of the mentioned terminating services arrive, or that the UE becomes active by itself, wishing to establish a voice call. The architecture was presented in sub-section 3.5.3, here we discuss the principal behavior of the UE and the network.

As long as CS fallback is enabled for a UE, MME and MSC/VLR keep an association for it via the SGs interface. This enables the UE, with the support of the network, to trigger the fallback in E-UTRAN, leave E-UTRAN radio, activate the GERAN/UTRAN radio interface and perform signaling to continue the call or Supplementary Service handling. Note that for SMS transfer special conditions apply, see below.

Apart from SGs interfaces being configured between VLR/MSC and MME nodes, and some data being present in MME for mapping locations, a precondition for this feature is that the two RANs (E-UTRAN and GERAN/UTRAN) overlap. Additionally, the UE has to perform the so-called "combined" procedures in NAS signaling with MME (for attach, TAU, detach); in this way the coordination between EMM (mobility states in E-UTRAN) and MM (mobility states in CS GERAN/UTRAN), including the creation and maintenance of an SGs association with MSC/VLR is achieved

☞ 3GPP TS 23.272 [29] contains the stage 2 description of CSFB.

Originating voice call

There are two variants, with PS handover support, which is shown on the overview level in Figure 4-38, and without PS handover support. In the first case the UE, being in E-UTRAN, issues a request for the MO call to the MME with an indication that fallback to the legacy CS domain is required, using the EMM protocol stack. After that the UE will switch its radio interface to the legacy system (GERAN or UTRAN) and perform an active mode PS handover, or ; after that, it issues another service request, now towards MSC/VLR and utilizing the MM protocol stack. The procedure is almost the same for idle or active mode UEs, as the "Service Request" brings the UE anyway into active mode. If there is no support for PS handover, ongoing parallel PS services need to be suspended for the duration of the CS call.

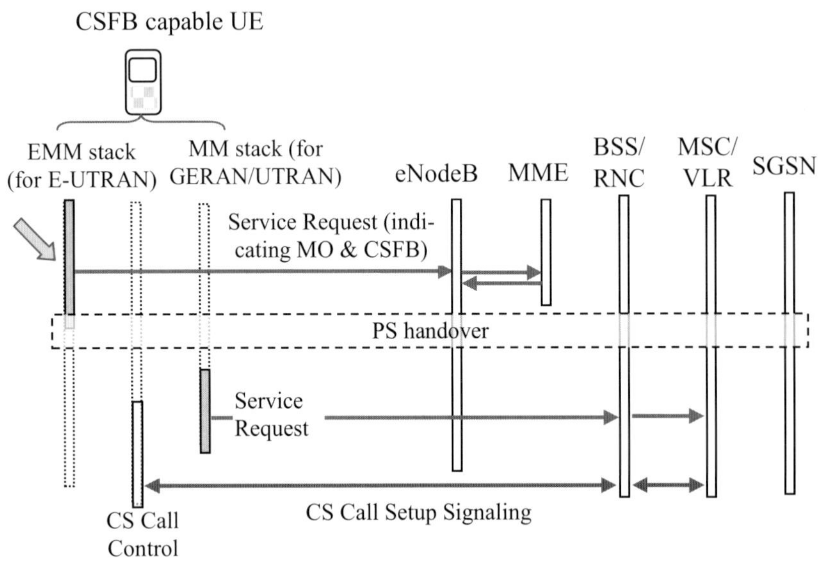

Figure 4-38: overview of originating voice call with CS fallback to GERAN/UTRAN (PS handover supported)

Terminating voice call (from CS domain)

The call arrives in the CS domain via GMSC at the visited MSC/VLR, see Figure 4-39. Thanks to the coordination between MM and EMM protocol machines the MSC/VLR knows that the UE can be paged only in E-UTRAN (via MME and eNodeB), using the special relay of paging message on SGs. The details or PS mode handover are not shown here. If the UE is in active mode (i.e. transferring data in E-UTRAN, when the CS voice call comes in), there is no need for paging in E-UTRAN; instead, the existing connection and a specifically defined NAS signaling message is used to inform the UE. The user still has the chance to accept the call, reject it explicitly or just not answer it. The supplementary services for redirection (upon no answer from the user, or not reachable condition) are valid also in this case and are handled by the MSC/VLR, based on corresponding signaling messages from MME.

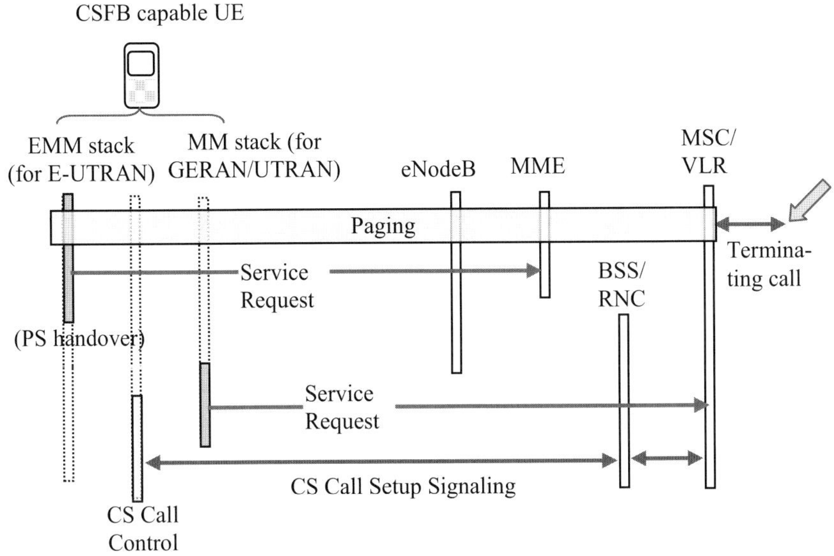

Figure 4-39: overview of terminating voice call in idle mode with CS fallback to GERAN/UTRAN

Network triggered Supplementary Services, like USSD and location requests, are supported by the same base mechanism; corresponding indications and all necessary data elements are defined for the paging message from MSC/VLR to MME and the onward NAS signaling to the UE.

Short Message Service

SMS is originally a CS based service, only recently 3GPP has defined a mechanism to transport SMS over generic IP access (SMS-IP); but if an operator does not want to move on to this deployment, she faces the problem of SMS transport via the Evolved 3GPP system (where a genuine "evolved" variant of legacy SMS delivery was not foreseen; see also sub-section 4.18 for a more complete overview of SMS handling in general). At least a feature in the NAS signaling protocol was developed which enables encapsulated SMS transport between MME and UE. In combination with the SGs interface realized for CS fallback for voice calls, the transport path between the legacy SMS infrastructure and the UE is complete:

SMS Service Center ←→ SMS Interworking MSC ←→ MSC/VLR ←→ MME ←→ UE

However, a real "fallback" to the legacy radio then does not need to take place, because the SMS is delivered already in the initial signaling phase, when UE is still in E-UTRAN.

The problem became more pressing at the very end of Rel. 8 standardization. Some operators wanted to use the SGs interface (between MSC/VLR and MME), originally designed for the signaling for CS fallback, **only** for SMS delivery, but **not** for the genuine fallback to CS for voice services. This would apply for the case that they deploy voice over IMS (over E-UTRAN), but not SMS over IP. Apart from some confusion in terminology resulting out of that, it also created the problem of standardizing the "SMS-only" behaviour of UE and the network, and of compatibility with the "full" CS fallback capable counterparts. The efforts of standardization of this feature were ongoing at the time of writing.

Selecting the proper domain for voice communication services

If both CS and PS with IMS are supported by the UE and available in the network, then there is still the question how a UE will select between them for voice communication services. Several factors are to be considered: network operator's preference, the terminal capabilities and user's preference. For the first one, configuration settings allow to control this selection:

1. use CS voice only,
2. use IMS PS voice only,
3. prefer CS Voice, with IMS PS voice as secondary, and
4. prefer IMS PS voice, with CS voice as secondary.

If an operator wants to dynamically configure it, the OMA DM protocol can be utilized (see sub-section 6.10), as these items were included in the definition of the more overall Managed Object for IMS.

4.15 Interworking with non-3GPP accesses

4.15.1 General

An early decision was made in the design for the Evolved 3GPP system: to realize mobility with non-3GPP access networks on the IP layer; the reasons were:

1. well established, IETF specified technology (client mobile IP for IPv4 and IPv6 were fully available, Proxy MIP development was already started and done in parallel);
2. independence of from 3GPP specific protocols;
3. Well suited to the all-IP vision, leveraging synergies in implementation, testing, deployment/operation.

It can be argued that more specialized procedures would increase performance, quality and efficiency during handovers; and indeed, for HRPD access of CDMA 2000 © such optimizations were additionally specified. Even on the IP layer some optimizations would be possible (see e.g. the concepts developed under the terms "low-latency" / "fast" handovers, for MIPv4 [30], [31], and MIPv6 [32]. But as a matter of fact, these have not been included in 3GPP Rel. 8 nor Rel. 9.

As stated with the architectural descriptions, PMIPv6, DSMIPv6 and MIPv4 FA mode are in general the options for mobility protocols for interworking with the Evolved 3GPP system. The latter two are client based (in IETF terminology also called host based), and the first one is a network based mobility scheme. With client based mobility, the UE changes its IP configuration (by acquiring a new, local IP address) on its own and does the mobility signaling, whereas with network based mobility the network takes care of it, and the UE does not change its IP address configuration (thus its IP layer remains unaware of the network internal mobility handling; however, its layer 2 implementation needs to support the movement of the IP address between different interfaces). The two mobility schemes are compared in Table 4-12 by a few main criteria.

Criterion	Client based mobility	NW based mobility
Impact on terminal implementation	high	low
Battery consumption	higher	lower
Mobility tunnel overhead on air interface	exists	does not exist
Dependence on access network capabilities	no	yes

Table 4-12: comparison of IP based mobility management schemes

The concepts of mobility schemes employed with non-3GPP accesses are described in subsequent sub-sections 4.15.2 to 4.15.4. Additional protocol details are presented in 6.2 and 6.3.

There are two special issues with non-3GPP access networks: trust detection and IP mobility mode selection, and these are elaborated in sub-sections 4.15.5 and 4.15.6.

4.15.2 Mobility based on Proxy Mobile IPv6 (on S2a and S2b)

Proxy Mobile IP (PMIP) was developed in parallel to SAE/LTE by IETF; in fact also an informal feedback loop (by interested companies and individuals) was installed between 3GPP and IETF working group.

PMIP originated from the NETLMM ("Network Based Local Mobility Management") effort in IETF; the target was to eliminate the need for mobility support in mobile clients by letting the network do everything necessary to enable IP layer mobility. This should have several advantages: less complexity in terminals, less overhead in data transmission on the capacity-critical radio interface, and also optimized signaling. On the other hand it requires support functionality in the network.

☞ The PMIPv6 base protocol is specified in IETF RFC 5213 [33]. Its usage by 3GPP including small extensions are described in 3GPP TS 29.275[34].

The fundamental concept of PMIP is that a node in the network emulates the home link for a client (in the following we use the 3GPP established term "UE" for such a client), and handles mobility related signaling and tunneling on behalf of it. This node is called "Mobile Access Gateway" (MAG); its function is derived from an access router: for those UEs which are authorized for network based mobility handling with PMIP it issues router advertisements containing the home network prefix. It detects movements of the UE to and from the access link (the link between UE and MAG). As shown in more detail in Figure 4-40, it keeps contact with an anchor entity in the home network ("Local Mobility Anchor", LMA). Despite the word "local" in this name of the entity, the way how it is used here enables global mobility for a UE.

Note: the term "home network" is used here in a different meaning than in subsection 3.10 (i.e. not "local network, connected at the home nodeB", but "home in MIP sense").

The general IETF terms for PMIPv6 functions can be mapped to the nodes in the 3GPP architecture as follows:

PMIP function	Realized by	Case	
MAG	Serving GW	1	PMIP based S5/S8
	ePDG	2	access to EPC from untrusted non-3GPP access network
	a node (Access Gateway) in a trusted non-3GPP access network	3	access to EPC from trusted non-3GPP access network
LMA	PDN GW	4	all except case 5
	Serving GW (in; only parts of LMA functions are required)	5	chained S2

Table 4-38: mapping of generic PMIPv6 functions onto 3GPP/SAE related ones

Ignoring the chained case for S2 for the moment, LMA is realized in EPC by the PDN GW. It connects to the PDN(s) and hosts the IP address for the UE. The PMIPv6 tunnel is realized by a binding of UE's PDN connection with the MAG's address and GRE downlink key, stored at LMA in the so-called "Binding Cache", and by a corresponding entries stored on MAG's "Binding Update" list. GRE encapsulation is employed due to the need for traffic separation in case of multiple PDN connections via the same PDN GW (=LMA). The MAG hosts the Proxy-Care-of-Address for the UE, which appears to the LMA as the CoA of the UE. Bindings are created, maintained and torn down by PMIPv6 signaling messages exchanged between MAG and LMA. Thus, the MAG function is largely equivalent to a "Home Agent" in client based MIP (see next sub-section).

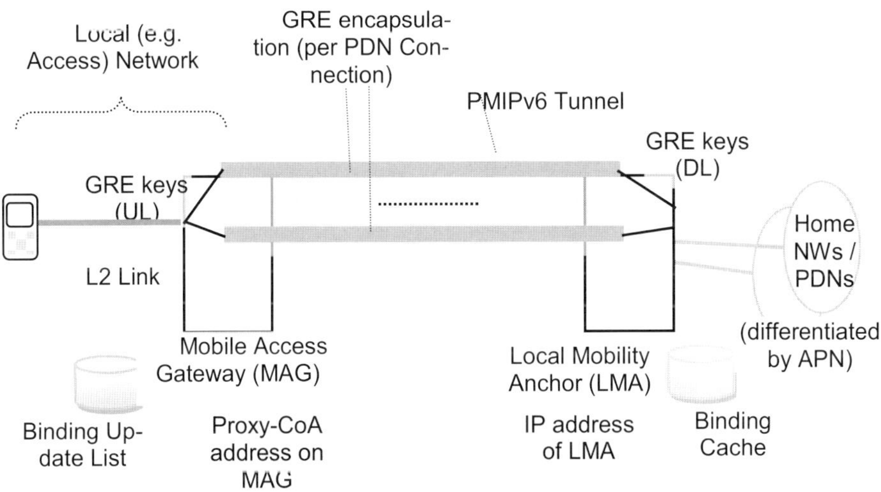

Figure 4-40: PMIPv6 tunneling concept

In contrast to GTP, the original PMIP has no notion of a "bearer"; in other words, it is not capable of separating traffic appropriately according to QoS or charging needs. If the difference is also reflected in separate IP addresses, this is not yet a problem. But it is also required to support identical IPv4 addresses, which may happen due to overlapping IP address space of private PDNs. In order to provide equivalent means as GTP, PMIPv6 usage in 3GPP relies on separation by additional encapsulation. GRE encapsulation is now mandatory for the user plane packets; GRE tunnels with different keys for up and downlink, per PDN connection and per interface (S5, S8, S2a, S2b). Additionally the whole concept of signaling of policy and charging information in parallel to the BBERF had to be invented (see sub-section 4.12).

PMIPv6 protocol stacks for user and control planes are depicted in Figure 4-41; an extension of PMIPv6 allows to use also IPv4 transport between MAG and LMA (which enables deployments with IPv4-only access networks), see subsection 6.2.1.

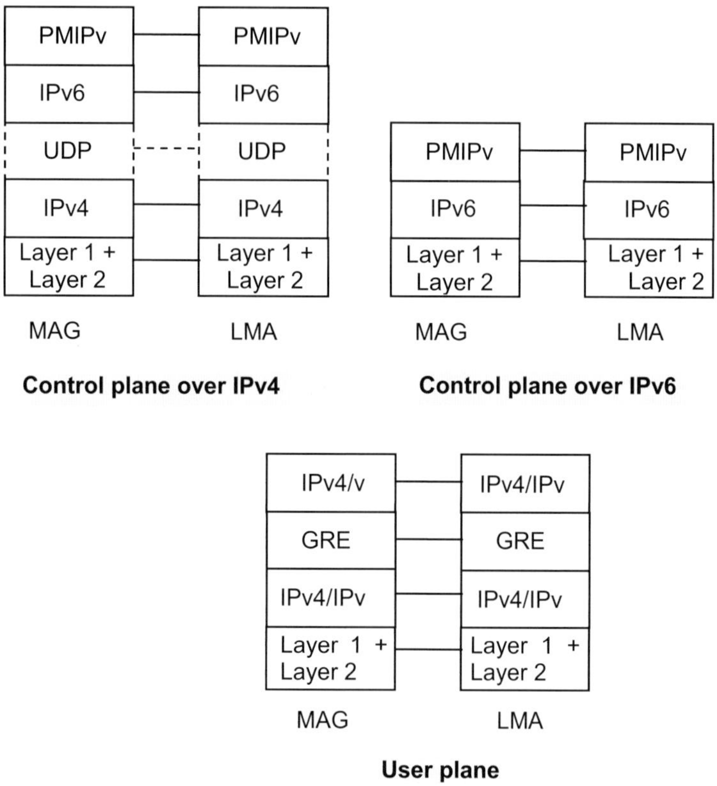

Figure 4-41: protocol stacks for PMIPv6

4.15.3 Mobility based on Dual Stack Mobile IPv6 on S2c

Dual Stack Mobile IPv6 (DSMIPv6, [35]) is an enhancement of Mobile IPv6 [37]; it belongs to the category of host-based mobility protocols. The concept is visualized in Figure 4-42.

For an understanding of the basic function the dual stack aspect (namely, that the local IP connectivity may be IPv4 or IPv6) can be ignored for the moment, and we explain it first for the IPv6 case.

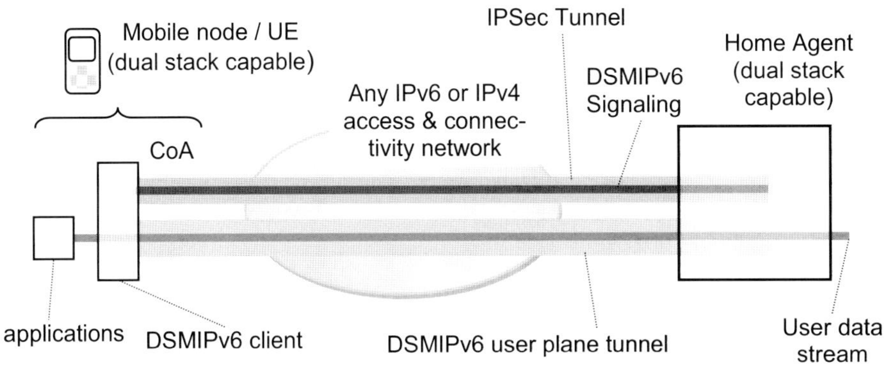

Figure 4-42: DSMIPv6 concept

The DSMIPv6 client of the UE (while being away from its home link and requiring mobility service) registers UE's locally acquired address as a Care-of-Address at the Home Agent (HA) for a binding with its IPv6 Home Address. The HA intercepts incoming IP packets and forwards them in an IPv6 mobility tunnel to the CoA. In the uplink direction the IP packets are equally first tunneled to the HA for decapsulation, before the normal routing takes over. The moving UE will update the binding in the HA with every change of local access network by DSMIPv6 signaling; within untrusted environements this signaling must be protected, and is then therefore tunneled via IPSec, in order to exclude security issues (e.g. redirection attacks, which would result in loss of connectivity for the target UE).

Before the UE can enjoy the DSMIPv6 mobility service, a bootstrapping procedure consisting of 6 phases has to be performed (see their explanation in Figure 4-43).

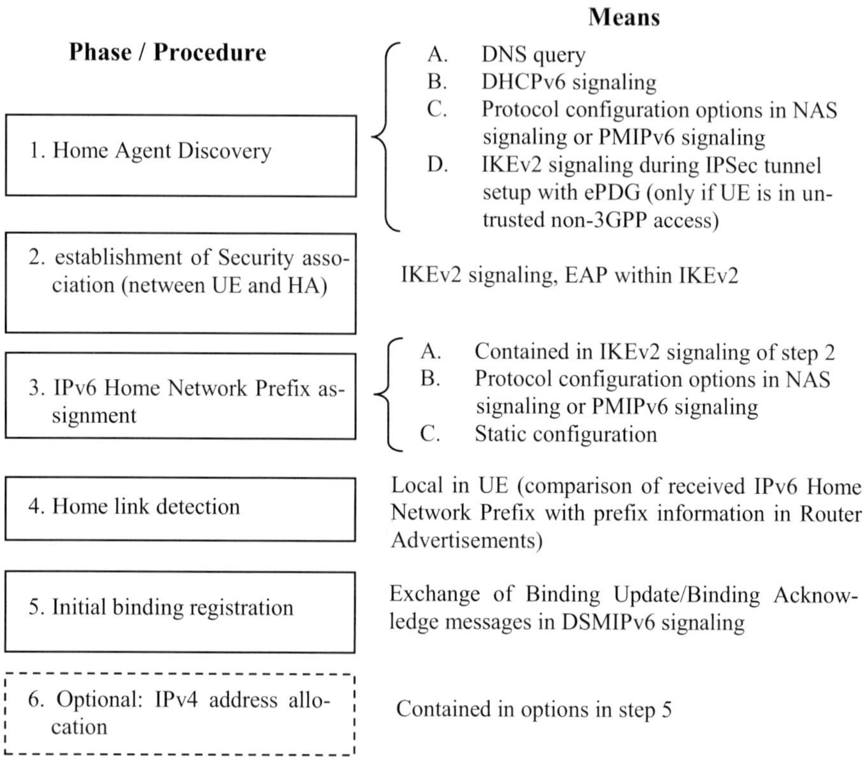

Figure 4-43: bootstrapping in DSMIPv6

In some cases it is necessary to change the discovered HA, e.g. if it is under high load or topologically not favourable. Therefore a HA re-allocation procedure is available. During the IKEv2 signaling phase 2, a REDIRECT indication is sent from the originally targeted HA to the UE, together with the new HA address. Then, the UE will stop the running IKEv2 signaling and start a new one with the new HA.

4.15.4 Mobile based on MIPv4 (FA mode) on S2a

Mobile IPv4 is an IPv4 based mobility scheme; there are two variants, co-located Care-of-Address (CoA) and FA CoA (see IETF RFC 3344 [36]). In the Evolved 3GPP system only the latter is used, and this description is limited to it. The motivation is that only FA mode gives operators a benefit with respect to IPv4 address space limitations.

Similar to MIPv6, a Home Agent (HA) acts as the anchor for a mobility tunnel to and from the mobile hosts/UEs. In the access network a Foreign Agent takes care of the mobility signaling with the HA. The UE runs a MIPv4 client; it discovers the FA and requests from it a mobility binding at the Home Agent on its behalf. The MIPv4 overall scheme is depicted in Figure 4-44.

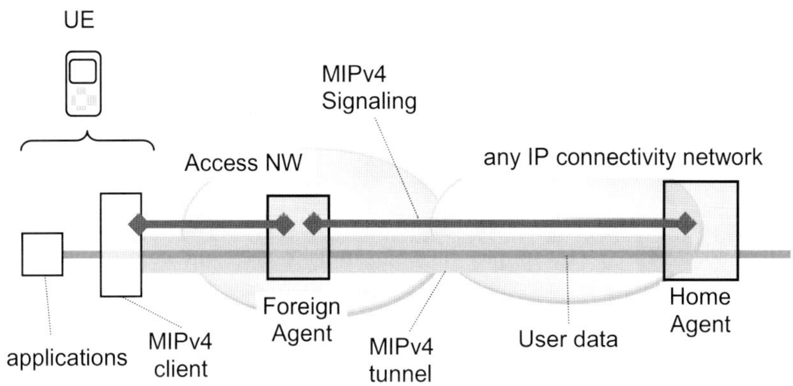

Figure 4-44: Mobile IPv4 (FA mode) concept

Within the Evolved 3GPP system, MIPv4 FA mode is used only in trusted non-3GPP access networks (for comparison, it is also used in 3GPP2's network architecture and WIMAX network specifications).

These steps are typically performed by UE and FA:

- FA discovery: either the UE issues a MIPv4 Agent Solicitation message, or the FA issues a MIPv4 Agent Advertisement message after detecting a new UE on its link.
- Initial Registration: the UE sends a MIPv4 REGISTRATION REQUEST to the FA, who forwards it to the HA. The NAI is used for identification of the UE. The address of the HA may either come from the UE (e.g. if statically configured), or the FA knows the HA address from its own configuration. As a result of the registration, a Registration Reply message is sent back to the UE.
- Re-registration: due to finite lifetime of the granted mobility bindings, the UE has to regularly initiate re-registrations towards the FA (who forwards them to the HA).
- Handover: the UE can detect movement from agent advertisements, or from layer 1 or 2 indications. In this case it sends a Registration Request with the new FA address.
- Deregistration: if the UE returns to the home link (or does not need mobility service for some reason) it can send a Registration Request message with lifetime 0; if sent from the home link, it will not pass a FA.

Security of MIPv4 signaling must be ensured; for that purpose authentication extensions already defined with the base MIPv4 protocol are used. The bootstrapping of MIPv4 security keys in the UE and correspondingly in the network (AAA server) with distribution to the mobility agents (PDN GW/HA and trusted non-3GPP access/FA) is done during access (re-)authentication. Two keys are used by the UE: one for the security relation between UE and FA, and one for the security relation between UE and HA.

4.15.5 Trust Detection

How does the UE get to know the trust relationship of the access network it is about to access EPC? The simplest mechanism is to use static configuration in the UE. The more flexible mechanism is a dynamic discovery. This is defined in a 3GPP specific extension parameter for the EAP-AKA/EAP-AKA' authentication / authorization methods. If the UE lacks static configuration data about a non-3GPP access network and does not receive a dynamic indication, it assumes that the non-3GPP access network is untrusted.

There is one problematic case: if the UE attaches to an access network where no 3GPP based authentication / authorization is run, it is not able to receive a dynamic trust indication. In this case it would start IKEv2 signaling for establishment of the IPSec tunnel, as it is required for the untrusted case. However, the non-3GPP access network could actually still be considered as trusted by the EPC operator, in which case the IPSec tunnel would be inefficient and unnecessary. At the time of writing there was no definite conclusion if this case should be optimized, e.g. by a network indication from ePDG to the UE to fall back to the trusted mode (thus stopping further IKEv2 signaling).

4.15.6 IP Mobility Mode Selection

Up to three "parties" have to be considered for the choice of the applicable one in a concrete case with their capabilities and preferences: 1) the UE, 2) the access network and 3) the EPC; however, EPC can take into account the access network capabilities in its decision. This leads to many combinations and the need for dynamic negotiation, which is done during the authentication/authorization phase.

Another decision is whether the allocated IP address should be preserved at all during the handover. This is relevant only if PMIPv6, as a network based mobility protocol, is used; with DMSIPv6 or MIPv4 the decision is with the UE as the mobility client. The UE requests for IP address preservation by setting the "attach type" to "handover" in the attach request (which is an integral part of the handover procedure); the network then may grant the preservation of IP address.

With all these diverse options for mobility protocols the problem arises how to establish a common understanding between the UE and the network. This is the task of IP mobility mode selection (IPMS); IPMS has to be run at initial attach in, and handover to, a non-3GPP access. The following table states the IPMS rules in detail:

Case	Capability/Preference of ...		Result
	UE	EPC	
1	DSMIPv6 only	DSMIPv6	DSMIPv6 can be used, providing IP address preservation. Trusted non-3GPP access network or ePDG allocates local IP address (to be used as CoA).
2		PMIPv6	PMIPv6 is used (if trusted non-3GPP access network supports it). IP address preservation is handled as per PMIPv6 specification.
3	DSMIPv6 and PMIPv6	DSMIPv6	See case 1.
4		PMIPv6	PMIPv6 is used (if trusted non-3GPP access network supports). IP address preservation provided by S2a means.
5	MIPv4 only	MIPv4	Trusted non-3GPP access network allocates a FA CoA.
6	No indication	PMIPv6	PMIPv6 is used (if trusted non-3GPP access network supports it). IP address preservation is handled as per PMIPv6 specification.

Table 4-14: rules for dynamic indication of IPMS

The IPMS indication uses a 3GPP specific extension parameter for the EAP-AKA/EAP-AKA' authentication / authorization methods with the AAA server. The AAA server receives UE's IPMS indication and sends back the IPMS decision with another 3GPP defined extension parameter in EAP-AKA/EAP-AKA'. For the representation of IPMS in the authentication / authorization message flow, see sub-section 5.1.2

Dynamic indication of IPMS is not mandatory, and using static configuration only is an option. Though, in this case several possibilities for mismatch exist, according to Table 4-15.

	Network		
UE	PMIPv6	DSMIPv6	MIPv4
PMIPv6	OK.	Mismatch (no access to EPC). Operator may allow local IP access (e.g. to Internet); details are not specified.	Mismatch (no access to EPC). UE is not able with the FA.

DSMIPv6	Mismatch. UE may be able to access EPC, depending if HA function is offered by the network and UE is able to detect it is on the home link.	OK.	Mismatch (no access to EPC). UE is not able with the FA.
MIPv4	Mismatch (no access to EPC). No FA function supported in access network.	Mismatch (no access to EPC). No FA function supported in access network.	OK.

Table 4-15: mismatch cases for static IPMS configuration

4.15.7 Trusted Non-3GPP Access

The complete protocol stacks for control and user plane between UE and PDN GW, via the trusted non-3GPP access when utilizing S2b are given in Figure 4-45. This depiction is enhanced, compared to the official specification, with the "plain" IP based control plane, e.g. for DHCP signaling.

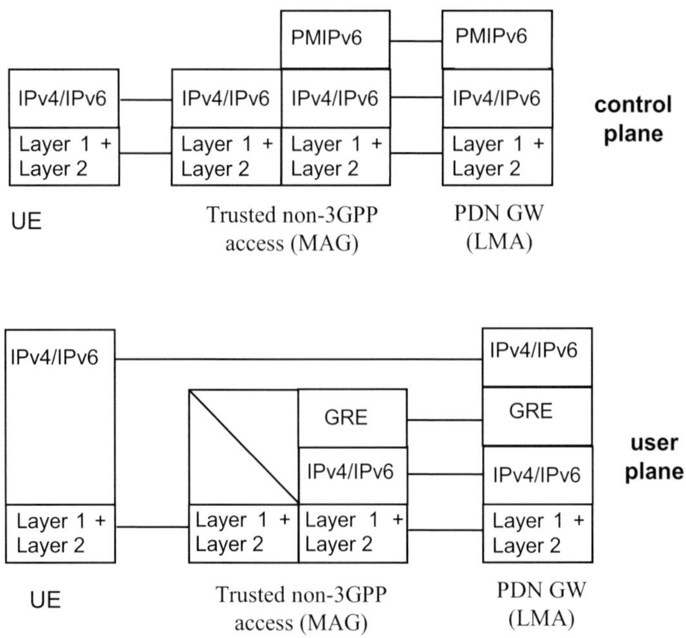

Figure 4-45: protocol stacks between UE and PDN GW (access from trusted non-3GPP access via S2a using PMIPv6)

For the IPv4 FA mode mobility scheme the protocol stacks look like given in-Figure 4-46. The MIPv4 specification [36] allows several tunneling mechanisms, e.g. IP in IP encapsulation.

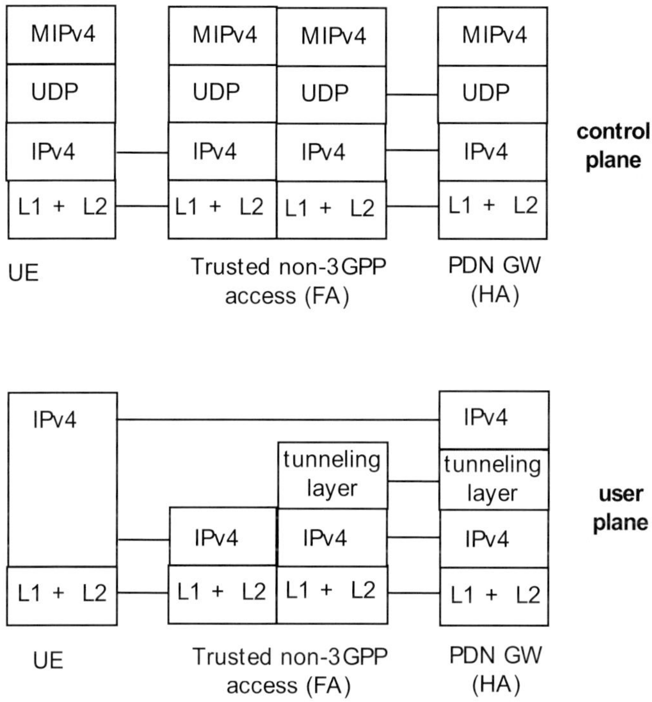

Figure 4-46: protocol stacks between UE and PDN GW (access from trusted non-3GPP access via S2a using MIPv4 FA mode)

The protocol stack for usage of DSMIPv6 over S2a is not shown graphically here, but it consists simply of:

- control plane: DSMIPv6 signaling directly between UE and PDN GW (functioning as HA), which runs on top of IPv4 or IPv6 (over any layers 1 and 2) between UE, trusted non-3GPP access and PDN GW;
- user plane: IPv4 or IPv6 directly between UE and PDN GW, tunneled over IPv4 or IPv6 between UE, trusted non-3GPP access and PDN GW.

That means, no dedicated security protection is required, as a consequence of the trust property with this non-3GPP access seen by the EPC.

4.15.8 Untrusted Non-3GPP Access

There are two options, the UE can either access the EPC via S2b (network based mobility) or via S2c (client based mobility). In both cases the local access router to which the UE is connected via L2 (the first IP node in the untrusted non-3GPP access network) is not shown. It is involved in the procedures (attach, detach, handover) only to the extent that the UE receives a local IP address, which is the prerequisite for using the higher layers (IKEv2 and IPSec).

The protocol stacks for control and user plane between UE and PDN GW across ePDG for access via S2b and S2c are shown in Figure 4-47 and Figure 4-48, respectively. With S2b/PMIPv6, PDN GW includes the LMA function, and ePDG the MAG function (see sub-section 6.2).

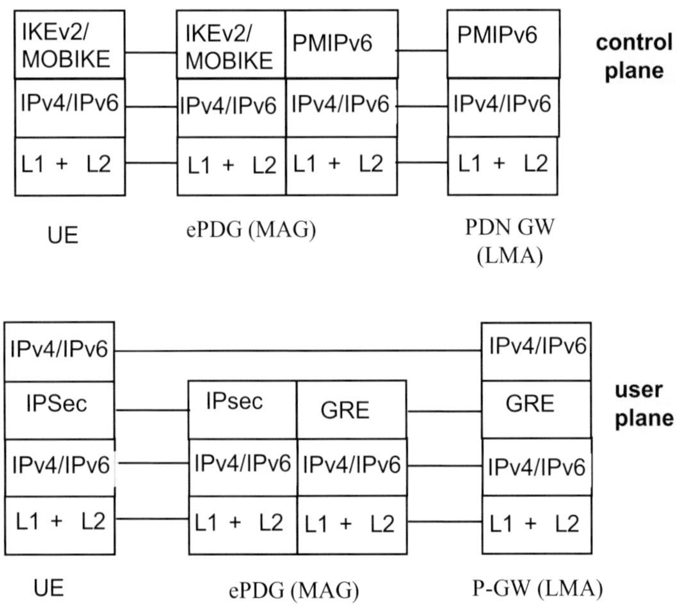

Figure 4-47: protocol stacks between UE and PDN GW (access via S2b)

As a result of the attachment to the EPC it receives then the IP address pertaining to the address pool of the PDN GW (linked to the APN and the PDN which is accessed via it).

For the S2c/DSMIPv6 case the depiction of the control plane is difficult, because IKEv2 signaling is needed in addition to (and, in the attach sequence, before) the DMSIPv6 signaling, which passes through the then established IPSec tunnel; protection of signaling is necessary in this case. With S2c the ePDG also assigns an IP address from its pool ("remote" IP address); it is used for DSMIPv6 signaling as the Care-of-Address. For DSMIPv6 the PDN GW includes the Home Agent function. Tunneling in the user plane layer is realized by IP encapsulation, as foreseen by the base MIPv6 specification [37].

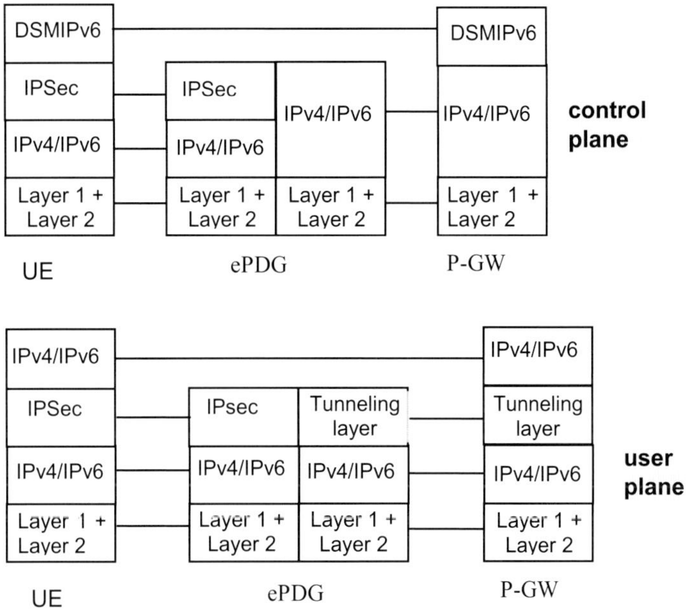

Figure 4-48: protocol stacks between UE and PDN GW for access via S2c

4.15.9 HRPD as an access network for EPC

For interworking with HRPD (this is the PS access in CDMA2000 © technology), optimized handover procedures have been developed (see architecture in sub-section 3.7 and message flow in sub-section 5.10.4). Figure 4-49 explains the rather simple protocol stacks for the S101 and S103 interfaces, which were already introduced in the architectural discussion in chapter 3. The control protocol S101-AP is realized on top of IP/UDP, and the user plane utilizes GRE encapsulation.

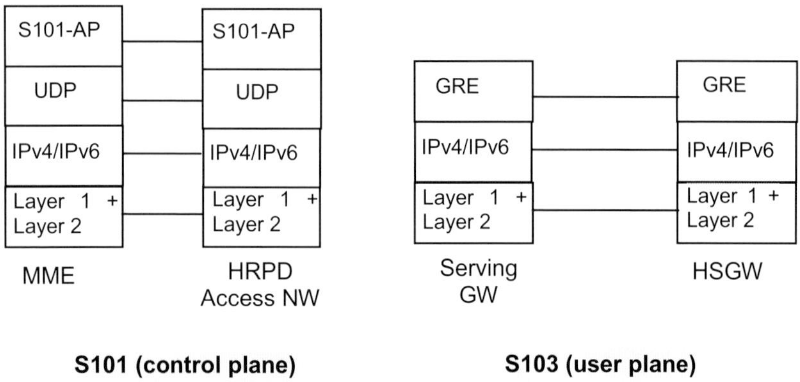

Figure 4-49: protocol stacks for S101 (left) and S103 (right) interfaces for optimized handover bet-ween HRPD and E-UTRAN

Apart from the specifically designed, optimized handover procedure, HRPD is generally treated as a trusted non-3GPP access.

4.15.10 Access Network Discovery & Selection Function (ANDSF)

Let us have a closer look how ANDSF, presented in chapter 2 as a functional entity in the architecture, can support a UE in the task of finding the access network for "best connectivity", which is also allowed by the operator.

ANDSF is designed to store and provide to the UE two categories of data (this is defined in an dedicated Managed Object within the OMA DM scheme):

1. Inter-system mobility policies: these are rules regarding what is allowed for the UE (by the operator) to select as access; rules have priorities associated with them, and only one rule can be active at maximum. The may include location/area data and time windows for their validity. The evaluation of rules results in preferred and restricted access technologies.
2. Access network discovery information: this data supports the UE in finding the most appropriate access NW. It contains, based on area/location data, per potential access technology the specific items to look for on the radio interface (e.g. SSID in case of WLAN, Network Access provider ID in case of WIMAX, PLMN/Tracking Area/cell identities in case of 3GPP access).

Additionally the location of the UE is also contained in the data structure, defined between UE and ANDSF (however, this information is "GET" only for ANDSF).

Several general observations can be made on ANDSF: firstly, it can be seen as a "dynamization" of the already existing, various lists for user and operator preferences as defined for I-WLAN (e.g. preferred WLAN list). Secondly, regarding the type of UE control by the network (or, seen from the other side, the degree of UE autonomy) the full range is possible – from fully UE centric to fully network centric. Thirdly, as a matter of fact, the agreement in 3GPP was also not to mandate a tight coupling of the communication between UE and ANDSF with handover events; on the other hand, it cannot be excluded that this communication occurs near to a handover. Forth, it was clearly decided that the access network selection for non-3GPP accesses shall not interfere with 3GPP's well established PLMN selection procedures.

The UE to ANDSF communication can be in push (ANDSF initiated) and in pull (UE initiated) mode. There is one potential problem, if policies need to be pushed down to the UE, but the UE has not yet had IP connectivity (this could apply if policies have changed for a UE currently in CS mode, and these policies should be available immediately), or if the UE has not yet discovered and made contact to the ANDSF for pull mode. In this case a triggering event like SMS would be used.

ANDSF discovery by the UE is either by static configuration, DNS query or DHCP query. For DNS query, a specific FQDN was defined by 3GPP in the form of "andsf.mnc<MNC>.mcc<MCC>.pub.3gppnetwork.org". (MNC and MCC are, as usual, the Mobile Network Code and the Mobile Country Code of the UE's HPLMN). If DHCP is used for ANDSF discovery, either an IP address or a domain name is given back. In the latter case, another resolution step by a DNS query is necessary.

The security of communication between UE and ANDSF can be guaranteed either by using the 3GPP defined Generic Bootstrapping Architecture (see 3GPP TS 33.402 [39]and TS 33.222 [40]), or by the OMA DM bootstrap and secure http solution.

An example how ANDSF data (access network discovery information) could be provided to a UE is given in Table 4-16. (For brevity we use the WIMAX related abbreviations NSP [Network Service Provider] and NAP [Network Access Provider], and WLAN related abbreviations SSID and BSSID). It is left to the reader to sketch the corresponding coverages of accesses.

ANDSF is a completely new concept for 3GPP, with no predecessor; despite the efforts spent for clarifying its functionality, at the time of writing it did still include some vagueness. E.g. the discussion was ongoing, whether at Rel. 8 UE with legacy I-WLAN capability at initial attach should always follow I-WLAN specific network discovery procedures first and only later take ANDSF policies into account, or if it could follow ANDSF policies from a previous download (at some time before this attachment); another source of confusion was the exact relation and difference in handling between a WLAN access to EPC and a WLAN access within the legacy I-WLAN.

UE's Location	AccessType = WiMAX	AccessType = WLAN
Cell 1	NSP-id 1: NAP-id 1, NAP-id 2 NSP-id 2: NAP-id 2, NAP-id 3	SSID = wlan1, BSSID = bs1 SSID = wlan2, BSSID = bs2
Cell 2	NSP-id 2: NAP-id 3	not available
Cell 3	not available	SSID = wlan1, BSSID = bs3 SSID = wlan4, BSSID = bs4
.....	....	....
Cell_n	NSP-id : NAP-id	SSID = wlan 7, BSSID = bs5

Table 4-16: example for ANDSF data (access network discovery information)

4.16 I-WLAN Mobility

In parallel to the design of the Evolved 3GPP system in the form of LTE/EPC, an enhancement of GPRS and I-WLAN for mobility was done; the complete network arrangements of these two PS technologies (up to GGSN in case of GPRS, and up to PDG in case of I-WLAN) are considered as "access systems", and an overlay system is introduced by placing a DSMIPv6 Home Agent logically "behind" these points of IP interconnection.

The network operator can choose either the GPRS access or the I-WLAN or none of these to serve as a home link in DSMIPv6 sense. In the former two cases the Home Agent would then be co-located with the GGSN or PDG (see upper part of Figure 4-50), respectively; in the latter case the Home Agent is separate from both of them (see lower part of Figure 4-50).

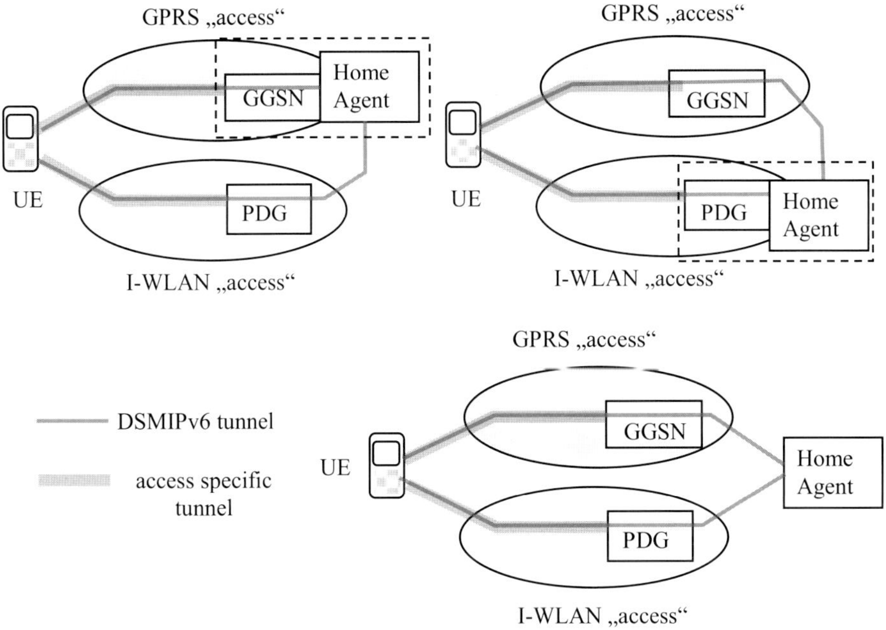

Figure 4-50: concept of I-WLAN mobility

4.17 Service continuity on IMS level

With the mechanisms described above, mobility between various PS access systems integrated under the umbrella of the Evolved 3GPP system/EPS (e.g. GPRS, E-UTRAN, WLAN, WIMAX) is possible with service continuity realized on the IP layer or below. However, PS access systems may remain separate from EPS still for some longer time, e.g. standalone GPRS, I-WLAN, TISPAN fixed network access or packet cable access, and converge only on the control layer (IMS). 3GPP's Rel. 8 therefore supports IMS based means for service continuity between different IP Connectivity Networks (IP-CANs); for these scenarios we use the term "access transfer", in order to avoid confusion with other scenarios of service continuity in the context of inter-device transfer. Under the assumption that the IP address changes with the IP-CAN, the concept and procedures described below are mainly suitable for real time communication based on UDP (TCP sessions would break).

The solution for IMS based service continuity foresees an SCC (Service Centralization and Continuity) application server (AS); it handles the leg specific signaling (preparation of new leg, continuity procedure and release of the old access leg). The specific functionality for IMS centralized services is required if service continuity between CS and PS should also be supported, but this is not in scope here. Figure 4-51 ignores irrelevant details (P-CSCF and any nodes for interconnecting networks) and shows that the UE has a control plane session ongoing with the IMS (represented by S-CSCF, SCC AS and optionally other application servers) via an IP-Connectivity Network A (IP-CAN A, e.g. GPRS), from where signaling to the remote party is performed; the 'media' (e.g. voice or multimedia streams) are exchanged via the same IP-CAN directly with the remote party. As the UE moves on to another IP-CAN B (e.g. I-WLAN), it is necessary to transfer ongoing IMS session(s) to the other access.

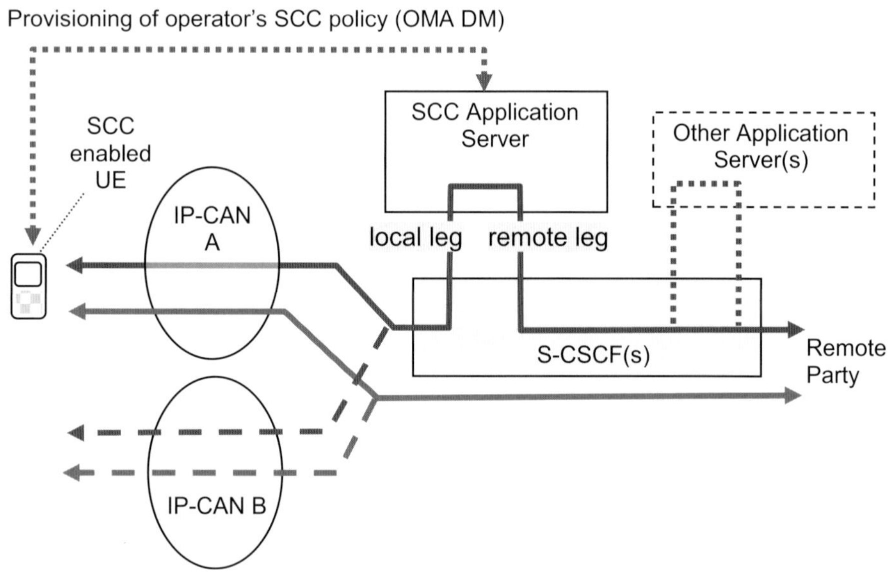

Figure 4-51: concept of service continuity on IMS level

Some policies are needed in the UE for an effective and optimal control of SCC; these are transferred according to a 3GPP defined Managed Object via OMA DM from the SCC AS to the UE.

In order to realize SCC for the UE on the local leg, the SCC AS anchors the IMS session. I.e. when an outward SIP INVITE is sent, it is detected by the so-called "Filter Criteria" and forwarded to the SCC AS, which completes the signalling to the remote leg (possibly via other Application Servers). For the terminating session the same anchoring is done, with the SCC AS being the last AS in the signaling path.

A prerequisite for the access transfer procedure, as presented here, is that UE and IMS support multiple simultaneous registrations; the UE also has to be able to establish connections with the two IP-CANs in question and maintain them for the duration of the access transfer. On the overall level, the following steps are needed:

1. The UE has an established the media path over the old IP-CAN.
2. The UE connects to the new IP-CAN and receives a new IP address, which is to be used for signaling and media.
3. The UE registers to the S-CSCF over the new IP-CAN.
4. The UE sends the trigger message access transfer to the SCC AS via the new IP-CAN, indicating the sessions to be transferred.
5. The SCC AS does the update signaling with the remote leg, and subsequently with the local leg (setup on the new leg and release on the old leg). This step requires only standard IMS procedures for 3rd party call control.

The message flow is explained in sub-section 5.14, including more details on the used parameters.

Based on the same SCC AS concept, extensions are being made for inter-device transfer in Rel. 9 and 10 (see also sub-section 4.17). The goal is to transfer either parts of a session (i.e. selected media components) or the whole session between two UEs. In the first case, a so-called "collaborative session" has to be established between the UE keeping the control, and the subordinate, "controllee" UEs on the local side and the SCC AS. In the second case also the control is handed over to the target UE (and no collaborative session is needed). The primary usage scenario is that a user has different, specialized devices at her disposal, e.g. apart from the mobile phone a large video screen or a high quality audio system; she can distribute media during ongoing sessions in order to optimize her service experience. Transfer of media to distant devices, e.g. for coordinated usage between family members, are also possible. Again, inter-device transfer is restricted to UEs under the same subscription (yet they may even roam in different networks). In Rel. 9 always only one controlling UE is allowed; this restriction will be relaxed in Rel. 10, to allow a more flexible usage.

The concept of collaborative session is explained in Figure 4-52. Technically speaking, SCC AS performs 3rd party call/session control, and acts as the concentrator and inspector of control signaling (SIP including SDP); however, the media streams still extend directly between the individual, local UEs and the remote party. The SCC AS handles the control signaling with the controllee UEs on behalve of the controlling UE, and the update to the remote party. The exact details of signaling are were still under debate at the time of writing.

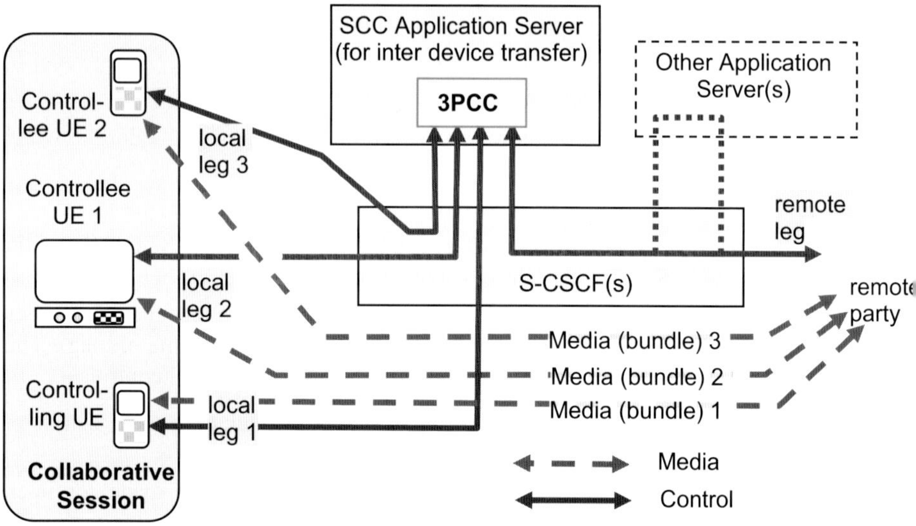

Figure 4-52: IMS based service continuity with inter-device transfer (preview on Rel. 9)

It may happen that the controling UE transfers all media to controllee UEs, and just keeps the collaborative session control.

☞ The functional description of IMS based service continuity (access and inter-UE transfer) is found in [41].

4.18 SMS handling

Short Message Service (SMS) was originally an unexpectedly successful by-product of signaling in the circuit switched mobile (GSM) network. It was later extended to the PS (GPRS) domain. It is not only a user service, but used also by operators for Over-the-Air provisioning (OTA).

Due to the introduction of additional access variants towards the 3GPP system and support of IP-CANs outside 3GPP's scope, like I-WLAN, a generic mechanism for interworking SMS via a PS/IP domain was specified. It consists of transporting SMS user contents and SMS status messages via IMS (encapsulated within a SIP message); the interfacing node is IP-SM-GW, which interfaces with the legacy SMS related nodes (SMS-Gateway-MSC or SMS-Interworking-MSC).

☞ SMS over IP is specified in 3GPP TS 23.204 [41]and 3GPP TS 24.341 [43].

With the advent of E-UTRAN, 3GPP decided not to develop another access specific extension of the original SMS concept, but first of all to apply the generic interworking mechanism (SMS over IP), with E-UTRAN/EPC being the IP-CAN; further, SMS transport is available in connection with the inter-domain signaling connection realized for CS fallback, see sub-section 4.14.3. In this case the originating or terminating SMS is routed via MME to/from MSC/VLR, where it enters/leaves the standard handling; in contrast to CS call handling, no actual change of the access (fallback) is necessary. Some specific enhancements were necessary in the signaling:

1. encapsulation mode in NAS signaling (between MME and UE): the NAS signaling message pair Downlink/Uplink NAS Transport was defined for that purpose.
2. encapsulation of SMS on the SGs interface (between MSC/VLR and MME): a message pair DOWNLINK- / UPLINK-UNITDATA is available in the SGs protocol stack.

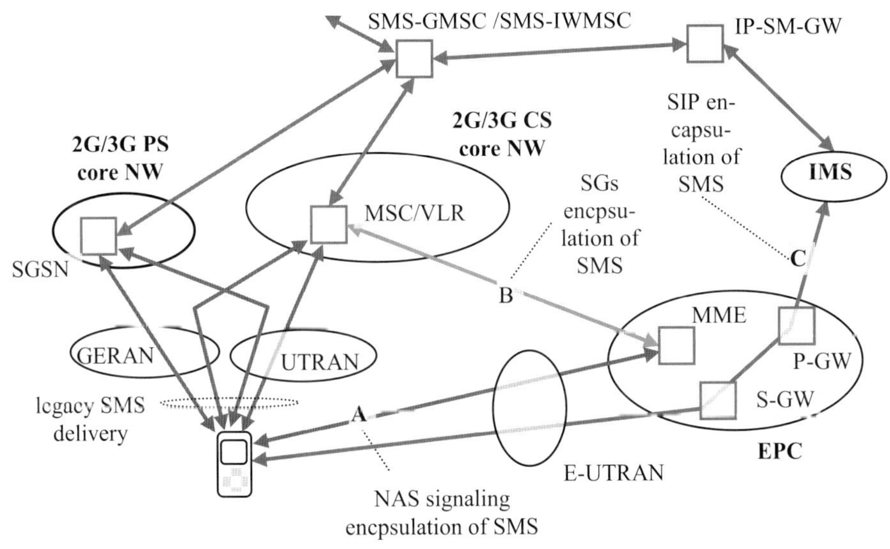

Figure 4-53: SMS handling in comparison: standard/legacy (A), utilizing domain interconnection for CS fallback (B) and over IP (C)

At the time of writing there was still a discussion ongoing in 3GPP, whether these means are sufficient; limitations in deployment (e.g. in roaming) were claimed, and several operators still thought that a generic SMS support over EPC/E-UTRAN would be required.

4.19 Closed Subscriber Groups

The concept of Closed Subscriber Group realizes special support by particular Home (e)NodeBs for defined members (in terms of UEs). The membership is expressed by an Allowed List maintained in the UE and in the network. The source of membership may be due to operator's or Home (e)NodeB owner's administration activity; in the latter case dertainly a secure front end system to the HSS is needed (but this is not standardized).

The main purpose of being a member of CSG is probably to take advantage of special billing/charging (but it is up to the operator to realize and market that). Another aspect is the automatically granted preferential access to the corresponding cells, with the benefit of (controlled) home cell resources. With the advent of home network based services, the CSG will also be a means to control access. The user should be aware of the CSG feature and usage, therefore the CSG identity, the Home NodeB name and the CSG type may be displayed on the UE.

The CSG membership may be only temporary; the use case here is that a home cell owner may wish to enable access for guests, on a temporary basis. At expiry of the membership, any service via a home cell pertaining to the CSG must be stopped. The storage of CSG in is in the UE, but additionally in the USIM (UICC), so that it is linked with the subscription and not only the terminal device.

In idle mode, additionally to the procedures of PLMN selection and cell re-selection/reselection, CSG selection comes into play. Therefore we refine here the picture given in sub-section 4.9.3. Although scenarios exist where the selection of a particular CSG may have higher priority than the currently registered PLMN, for Rel. 8 it was decided to decouple and serialize PLMN and CSG selection. So always a PLMN is selected first and then the CSGs available within that PLMN are detected and provided to the UE from the radio layers (e.g. for the purpose of displaying them to the user).

CSG in idle selection happens on two levels, on the NAS level and on the AS level. On NAS level it is possible in automatic or manual mode. There is a slight difference between these, as in manual one single CSG, out of the available ones, is selected for use. In automatic mode the list of available CSGs is used by the UE radio access network, and the selection is made based on measurements/signal quality. The overall process diagram is shown in Figure 4-54, with the typical sequence of actions indicated by numbers.

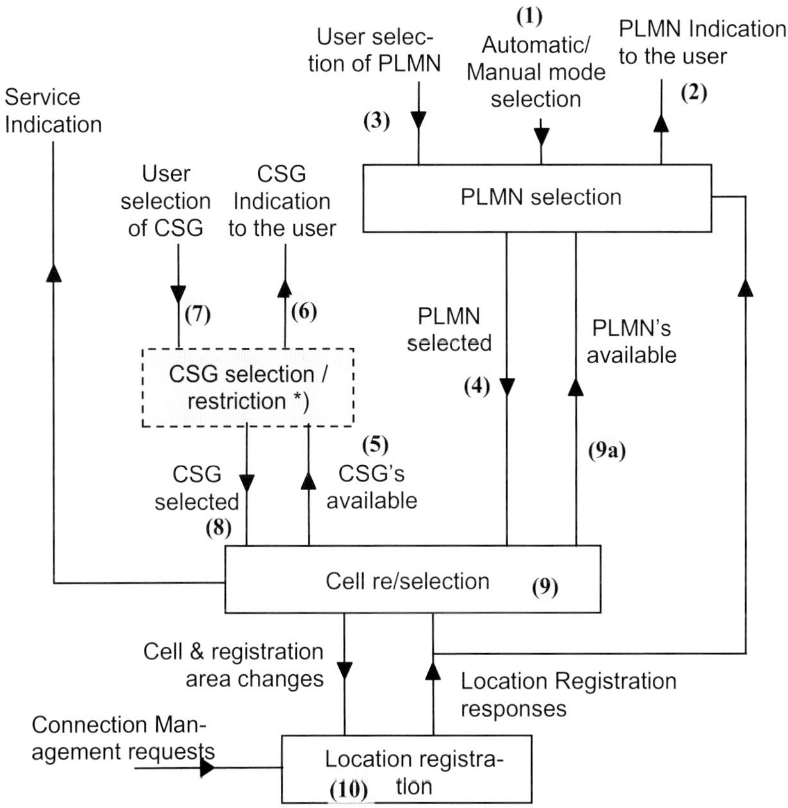

Figure 4-54: PLMN and CSG selection in idle mode

4.20 Multiple PDN Connectivity

The UE gains connectivity to one default or specifically adressed PDN/APN by performing the initial attach. With diverse and multiple applications running, a UE may need to connect to more than one PDN/APN at a time. The support of multiple PDN connectivity is optional for the UE.

This case of multiple PDN connectivity comes in two forms with a major difference, as visualized in Figure 4-55:

- The additional PDN/APN is accessible through the same, already allocated PDN GW, or
- The additional PDN/APN is only accessible via an additional PDN GW.

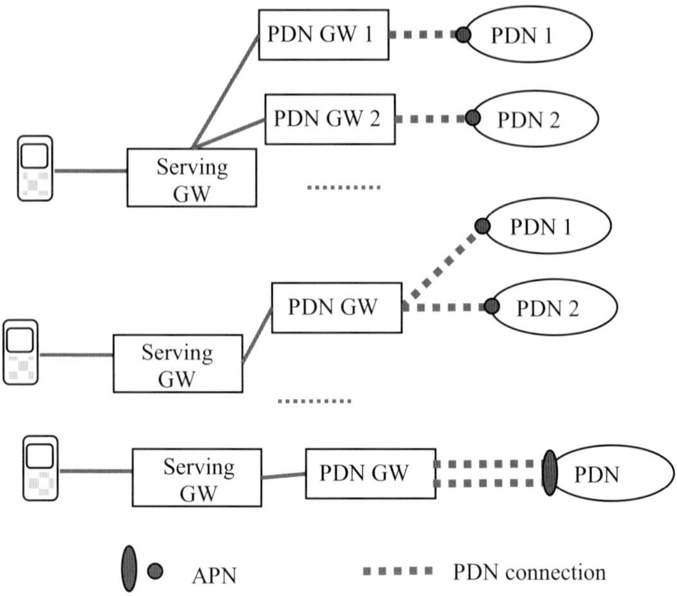

Figure 4-55: multiple PDN connectivity via different PDN GWs (upper part), via the same PDN GW through different APNs (middle part), and through the same APN (lower part)

In addition to the multiplicity of PDN GWs and APNs it is necessary to cater for multiple PDN connections through one and the same APN (lower part of Figure 4-55).

The situation is different for GTP and the rest of mobility protocols. Multiplicity of PDN connections is naturally supported by the GTP protocol, even for the same APN. Within Rel. 8 it was not possible to align all mobility procedures with it, and differences are accepted for the sake of simplicity. In practice it means that for multiple PDN connectivity via one and the same PDN GW from untrusted non-3GPP access, multiple IKEv2 sessions are used for mobility signaling and multiple IPSec tunnels for traffic between the UE and ePDG; in other words, the option to use one IKEv2 session and several child Security Associations with appropriate traffic selectors for IPsec tunnels was not taken on. Similarly, with DSMIPv6 also separate IKEv2 Security Associations are established between the UE and the HA, in case of multiple PDN connectivity via the same HA.

PMIPv6, together with the extension of GRE, is somewhere in between; GRE realizing traffic separation "inside" a PMIP tunnel, while mobility signaling is separated (i.e. there are no bulk mobility registrations; however, bulk binding revocations are used, based on recent developments in IETF, in the case of MAG or LMA failures). Multiple PDN connections for the same APN are not support ed via PMIPv6 in Rel. 8; this is subject to enhancements in Rel. 9.

In light of the above protocol issues, the following simplifications with multiple PDN connectivity were necessary in Rel. 8:

- the same mobility management scheme has to be used for each accessed PDN;
- handover within 3GPP access with a change of the S5/S8 protocol variant is not supported with multiple PDN connections via the same APN.
- at handover over from 3GPP access with GTP as EPC protocol to a non-3GPP access with PMIP (S2a or S2b): when the UE has currently more than one PDN connection to the same APN, only one PDN connection to the given APN will remain in the target non-3GPP access.

The latter two restrictions will be removed in Rel. 9; but even then still all PDN connections through one and the same APN need to be established via the same access network (which puts some restriction for handovers).

4.21 Identities

4.21.1 Network Identification

The identification of PLMNs by Mobile Country Code (MCC) and Mobile Network Code (MNC) remains in the Evolved 3GPP system as before. They appear as part of the full and permanent subscriber identification IMSI, as well as identifiers (FQDNs) of services (APNs) and network nodes; in the coding of the latter, a 2 digit MNC is always filled up to 3 digits by a leading zero.

4.21.2 UE Identification in 3GPP E-UTRAN Access

The identification of a UE is based on a globally unique MME identification (GUMMEI) and a so-called M-TMSI, see Figure 4-56; the structure of leading prefixes of identifiers is kept as in the legacy system, i.e. based on MCC and MNC. In order to allow more efficient radio procedures, e.g. for paging, a shortened temporary UE identifier is also defined (S-TMSI); however, this one is only unique within a MME group.

The MME identitification included therein consist of a MME Group id (8 bits, but coded with two bytes in hexadecimal representation) and MME id (16 bits, 4 bytes in hexadicimal representation).

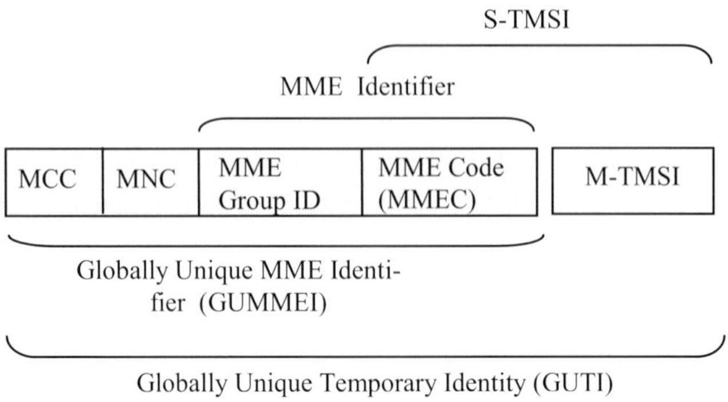

Figure 4-56: identities used for addressing UE and MME

If PMIP is used as EPC protocol, then a UE's identity is coded in the form of a NAI (see below).

4.21.3 Tracking Area Identity

Global unique identification of tracking areas is achieved by virtue of Tracking Area identities; their leading part contains MCC and MNC, while the Tracking Area Code remains as the unique part within one PLMN (see Figure 4-57).

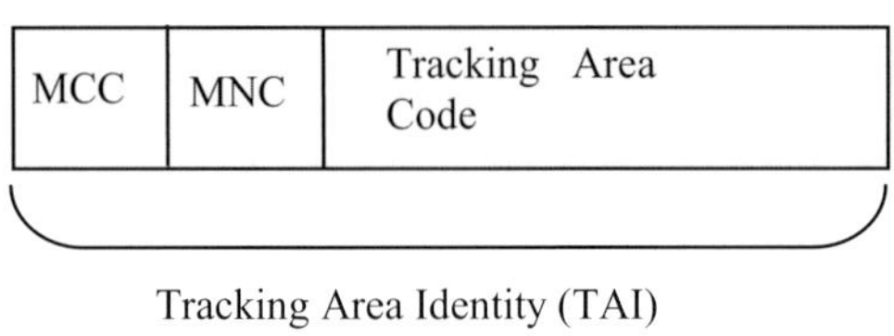

Figure 4-57: Tracking Area Identity

4.21.4 UE Identification for PMIP and in non-3GPP access

In the signaling procedures for access to the EPC (e.g. for authorization of service including mobility via interfaces S2a, S2b and S2c), the concept of the Network Access Identifier (NAI, see [44]) is used. Several variants of NAI exist:

- a "root NAI" is used in non-roaming and identifies the user and the HPLMN;

- a “decorated NAI” is used in roaming and identifies the user, the VPLMN and HPLMN;
- a “fast re-authentication NAI” is used in the special, shortened procedure for re-authentication;
- a “pseudonym” is used as a temporary identiy, comparable to (P)-TMSI in the legacy system.

The general format is “username@realm”, where ‘realm’ is a domain name and ‘username’ is a unique identification of the user therein. The IMSI is encoded within the username; a leading digit “0” is used for a NAI with EAP-AKA and a leading digit “6” for NAI with EAP-AKA’; this gives the following result for the root NAI:

“0<IMSI>@nai.epc.mnc<MNC>.mcc<MCC>.3gppnetwork.org”

As an example, if the IMSI is 232119876543210 (where the MCC part is 232, and the MCC part is 15), then the complete root NAI is

“0987654321023211@nai.epc.mnc232.mcc011.3gppnetwork.org”

If the UE has no USIM available (which is possible for access to emergency services, to come with Rel. 9) only the IMEI can be used to identify the UE. In this case the format is:

“imei<IMEI>@sos.invalid”

4.21.5 Identification of CSGs and Home NodeBs / Home eNodeBs

The CSG-ID identifies a collection of cells within E-UTRAN or UTRAN, where access is limited to particular subscribers (UEs). It is not intended for human use, so the CSG-ID is simply a 27-bit string. The length stems from the requirement to identify uniquely at least 125 million CSGs in one PLMN.

In contrast, the identification of a Home NodeB or Home eNodeB is to be displayed on the UE and read by the human user. It is therefore a name of 48 bytes length (coding in UTF-8 format). The Home NodeB/Home eNodeB name is not guaranteed to be unique in a PLMN.

4.21.6 Identification of a service (Access Point Name)

The concept of APNs had been developed within GPRS and is re-used also with EPC. The legacy APN (for GPRS) is resolved by DNS procedures to the IP address of a GGSN; it consists of network identifier and operator identifier, and ends with the label “gprs”, e.g.:

"internet.mnc011.mcc232.gprs"

Similarly, for usage within EPC, the APN-FQDN is defined as a more general name; it is resolved by DNS to the IP address(es) of a PDN GW. The characteristic label "apn.epc" is now inserted before between network identifier and operator identifier, and also the label "gprs" is substituted by "3gppnetwork.org". As a result, the above given GPRS specifc APN would be mapped to:

"internet.apn.epc.mnc011.mcc232.3gppnetwork.org".

APNs may be configured in the HSS (for default PDN connectivity) per user, or input from the UE in the procedures for attach or additional PDN connectivity (see sub-sections 5.4 and 5.5).

4.21.7 Identification of network nodes

Within node selection procedures, FQDNs are used and finally resolved by DNS requests to IP address(es). The FQDN format for an MME node is constructed as:

"mmec<MMEC>.mmegi<MMEGI>.mme.epc.mnc.<MNC>.mcc<MCC>.3gpp network.org",

where MMEC and MMEGI are the MME identifiers described above. If only an MME pool needs to be addressed, the following shorter FQDN format is used:

"mmegi<MMEGI>.mme.epc.mnc<MNC>.mcc<MCC>.3gppnetwork.org"

An ePDG FQDN consists of seven labels and has the following structure:

"epdg.epc.mnc<MNC>.mcc<MCC>.pub.3gppnetwork.org"

4.21.8 Access Network Identity

Trusted non-3GPP access networks are specifically identified, due to need for a separation of security domains; the Access Network identifier (ANID) enters the key derivation function of EAP-AKA' (see sub-section 5.1.2). Its general format/encoding is defined by 3GPP. It consists of a technology specific ANID prefix and possibly an additional string. Currently four ANID prefixes are defined (in brackets the standardization body responsible for assigning additional identifying strings is given):

- "HRPD": identifies an access network of HRPD cdma2000® technology (defined by 3GPP2);
- "WIMAX": identifies a WIMAX access network (WIMAX Forum);

- "WLAN": identifies a WLAN access network (IEEE 802);
- "ETHERNET": identifies a fixed access network (IEEE 802).

4.22 Feature Control

The variety of features with optional support both in the network and in the UE requires a systematic handling. For that purpose the network support information and UE capability information are used. NAS signaling messages ATTACH ACCEPT and TRACKING AREA ACCEPT are used to exchange this information between UE and the network.

Currently (i.e. in 3GPP Rel. 8) the EPS network feature support information contains only a flag indicating whether IMS voice over a PS session in S1 mode is supported or not; extensions for more features can be expected, as soon as they are standardized. The corresponding information element in NAS signaling messages for GPRS similarly contains the indication about IMS voice over PS session support in Iu mode, and additionally: whether (1) mobile originated Location Service requests (via the PS domain), and whether (2) MBMS is supported.

The mobile terminal/user equipment may vary broadly in the support of features. The network needs to know about that in order to be able to react accordingly. From history, such categorizing data is split into several items.

In the legacy system (pre-Rel. 8) the established terminology for a user's equipment is "Mobile Station" (MS) and accordingly the information element describing the terminal capabilities is termed MS network capability. It is signalled from the MS to the SGSN during the attachment and Routing Area Update and consists of bit coded information, describing e.g.

- whether one of the several GPRS Encryption Algorithms is supported or not,
- the support of services like SMS, Location Services and Supplementary Services,
- capability for PS inter-RAT handover to E-UTRAN S1 and UTRAN Iu mode,
- ISR support,
- CSFB support,
- SRVCC to GERAN/UTRAN capability,
- EPC capability (whether the MS supports access to the EPC via access networks other than GERAN or UTRAN; if only a SIM is supported, this must be set to "0" by the MS).

Additionally, the Mobile Station Classmark 1, 2 and 3 items contain information indicating the specification phase support, related to the radio capabilities (band usage, duplexing, channels, slots, coding, modulation, positioning, measurement, power management), type of RAN to core network interface support,

GSM encryption algorithm support, Voice Broadcast Service and Voice Group Call Service support. This data is included in several uplink signaling messages.

For the evolved system, the UE network capability information element was added. It contains the information concerning aspects of the UE related to EPS or interworking with GPRS. It is sent from the UE to the MME in uplink messages. The bits therein encode:

- whether a particular EPS encryption algorithm, EPS integrity protection algorithm, UMTS encryption algorithm and UMTS integrity protection algorithm (of the several possible ones per case) is supported,
- whether the UE has a preference of the GSM default alphabet over the UCS2 alphabet or not, and
- whether SRVCC from E-UTRAN to cdma2000® 1xCS is supported or not.

For the radio system it is also important to know about the basic capabilities of the mobile terminals, this can be used e.g. in handover decisions. Therefore specific classes of UEs have been defined:

Class	**Uplink**	**Downlink**
A	[50] Mbit per second	[100] Mbit per second
B	[25] Mbit per second	[50] Mbit per second
C	[2] Mbit per second	[2] Mbit per second

Table 4-17: UE classes in E-UTRAN

4.23 Overview of information storage

Various nodes of the EPS have to store static and dynamic data (the former is assumed to change only upon subscriber administration and O&M activities, like network [re-]configuration).

The principle is given in the following figure 4-58, considering mainly items specific to the EPS. UE Identifiers are not shown except in HSS and USIM, and for the sake of brevity and simplicity we also deviate slightly from the exact definitions of contexts. Also node addressing information is skipped here.

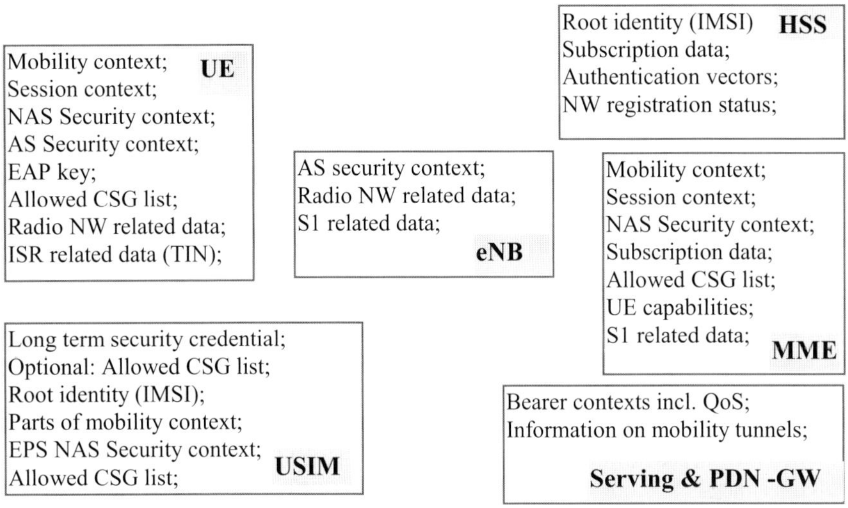

Figure 4-58: data storage for EPS (principle overview)

The UE stores this data (partially in non-volatile memory):

- mobility context: these are permanent and temporary identities, assigned TA list, forbidden TA and PLMN lists;
- session context: APNs in use together with their APN-AMBR, PDN type and IP address(es); bearers in use together with their QoS parameters and TFTs;
- NAS security context: it consists of a key (K_{ASME}) with the associated key set identifier, the UE security capabilities and selected cryptographic algorithms, and the uplink and downlink NAS COUNT values (see sub-section 4.8). If this is a "full" security context, it contains also the derived keys K_{NASint} (for NAS integrity protection) and K_{NASenc} (for NAS encryption).
- AS security context: consists of a security key and related parameters, selected AS level cryptographic algorithms and counters used for replay protection;
- EAP key: applies if authentication in the non-3GPP access has been performed;
- list of allowed CSGs: it optionally present, if CSGs are supported by the UE (see sub-section 4.19);
- TIN: indicates the type of temporary identity to be used in the next update to the network (ISR related, see sub-section 4.13);
- radio NW related data: system information, data for radio measurement, PLMN and cell selection.

The USIM, the 3GPP defined application for subscriber identification on the UICC ("smart card") inserted into the user equipment, has been evolved further in Rel. 8, according to the needs arising from EPS and additional features coming along with Rel. 8. The following EPS related data is stored, in addition to the long term security credential and permanent identity (IMSI):

- optionally: list of allowed CSGs (for storing Closed Subscriber Group membership information for home cells);
- selected EPS mobility management parameters: UE's Global Unique Temporary Identifier (GUTI), last visited TA and EPS update status are stored;
- EPS NAS security context;

The support of EPS in the USIM application toolbox (a platform for SW development based on USIM data) can be noted generally as an enhancement in Rel. 8; as an additional Rel. 8 feature, the Contact Manager, based on OMA's Data Synchronization [45] is available as an application on the USIM.

For the purpose of this principle overview it should be sufficient to state that an eNodeB needs to store radio NW related, S1 related data and the AS security context.

The HSS contains the subscription data, with the root identity (IMSI) being the primary key; on the overall level the important data to store are subscribed UE-AMBR (which limits the aggregate data transmission rate for non-GBR bearers over all PDN connections), access restrictions and overall subsribed charging characteristics. On the level of PDN connection the subscription data contains e.g. subscribed IP address, PDN type, APN, subscribed QoS, subscribed APN-AMBR, PDN specific charging characteristics and an indication whether PDN GW is allowed in VPLMN.

A list of data stored in MME and Serving GW is found in the Annex.

References

[1] 3GPP TS 36.211: "Evolved Universal Terrestrial Radio Access (E-UTRA); Physical Channels and Modulation"

[2] 3GPP TS 36.212: "Evolved Universal Terrestrial Radio Access (E-UTRA); Multiplexing and channel coding"

[3] 3GPP TS 36.213 "Evolved Universal Terrestrial Radio Access (E-UTRA); Physical layer procedures"

[4] 3GPP TS 36.214 "Evolved Universal Terrestrial Radio Access (E-UTRA); Physical layer – Measurements"

[5] H. Holma, A.Toskala (Editors): „LTE for UMTS – OFDMA and SC-FDMA Based Radio Access", John Wiley & Sons (2007)

[6] S. Sesia, M. Baker, I. Toufik (Editors): "LTE: The UMTS Long Term Evolution: From Theory to Practice", John Wiley & Sons (2009)

[7] 3GPP TS 36.321: "Evolved Universal Terrestrial Radio Access (E-UTRA); Medium Access Control (MAC) protocol specification"
[8] 3GPP TS 36.322: "Evolved Universal Terrestrial Radio Access (E-UTRA); Radio Link Control (RLC) protocol specification"
[9] 3GPP TS 36.323: "Evolved Universal Terrestrial Radio Access (E-UTRA); Packet Data Convergence Protocol (PDCP) specification"
[10] 3GPP TS 36.331: "Evolved Universal Terrestrial Radio Access (E-UTRA): Radio Resource Control (RRC); Protocol specification"
[11] IETF RFC 3075 (July 2001):"Robust Header Compression (ROHC): Framework and four profiles: RTP, UDP, ESP, and uncompressed"
[12] 3GPP TS 22.001: "Principles of circuit telecommunication services supported by a Public Land Mobile Network (PLMN)"
[13] 3GPP TS 22.002: "Circuit Bearer Services (BS) supported by a Public Land Mobile Network (PLMN)"
[14] 3GPP TS 29.273: "Evolved Packet System (EPS); 3GPP EPS AAA interfaces"
[15] 3GPP TS 33.210: "Network Domain Security; IP network layer security"
[16] IETF RFC 3748 (June 2004): "Extensible Authentication Protocol (EAP)"
[17] IETF RFC 4187 (January 2006): "Extensible Authentication Protocol Method for 3rd Generation Authentication and Key Agreement (EAP-AKA)"
[18] IETF RFC 5448 (May 2009): "Improved Extensible Authentication Protocol Method for 3rd Generation Authentication and Key Agreement (EAP-AKA')"
[19] 3GPP TS 24.301: "Non-Access-Stratum (NAS) protocol for Evolved Packet System (EPS); Stage 3"
[20] 3GPP TS 23.122: "Non-Access-Stratum (NAS) functions related to Mobile Station (MS) in idle mode"
[21] 3GPP TS 36.304: "Evolved Universal Terrestrial Radio Access (E-UTRA); User Equipment (UE) procedures in idle mode"
[22] 3GPP TS 22.011: "Service accessibility"
[23] 3GPP TS 23.107: "Quality of Service (QoS) concept and architecture"
[24] 3GPP TS 23.401: "GPRS enhancements for E-UTRAN access"
[25] 3GPP TS 23.125: "Overall high level functionality and architecture impacts of flow based charging; Stage 2"
[26] 3GPP TS 23.203: "Policy and charging control architecture"
[27] Volga Initiative, "VoLGA – Requirements V1.1.1 (2009-02)" and "V.o.L.G.A. Stage 2 V0.2.0 (2009-04-29)"
[28] 3GPP TS 23.216: "Single Radio Voice Call Continuity (SRVCC); Stage 2"
[29] 3GPP TS 23.272: "Circuit Switched Fallback in Evolved Packet System; Stage 2"
[30] IETF RFC 4881 (June 2007): "Low-Latency Handoffs in Mobile IPv4"
[31] IETF RFC 4988 (October 2007): "Mobile IPv4 Fast Handovers"
[32] IETF RFC 5268 (June 2008): "Mobile IPv6 Fast Handovers"
[33] IETF RFC 5213 (August 2008): "Proxy Mobile IPv6"
[34] 3GPP TS 29.275: "Proxy Mobile IPv6 (PMIPv6) based Mobility and Tunnelling protocols;Stage 3"
[35] IETF RFC 5454 (March 2009): "Dual-Stack Mobile IPv4"
[36] IETF RFC 3344 (August 2002): "IP Mobility Support for IPv4"
[37] IETF RFC 3775 (June 2004): "Mobility Support in IPv6"
[38] 3GPP TS 23.402: "GPRS architecture enhancements for non-3GPP accesses"

[39] 3GPP TS 33.402: "3GPP System Architecture Evolution (SAE): Security aspects of non-3GPP accesses"
[40] 3GPP TS 33.222: "Generic Authentication Architecture (GAA); Access to network application functions using Hypertext Transfer Protocol over Transport Layer Security (HTTPS)"
[41] 3GPP TS 23.237: "IP Multimedia Subsystem (IMS) Service Continuity; Stage 2"
[42] 3GPP TS 23.204: "Support of Short Message Service (SMS) over generic 3GPP Internet Protocol (IP) access; Stage 2"
[43] 3GPP TS 24.341: "Support of SMS over IP networks; Stage 3"
[44] IETF RFC 4282 (December 2005): "The Network Access Identifier"
[45] http://member.openmobilealliance.org/ftp/Public_documents/DS/Permanent_documents/

Chapter 5: Functions and Procedures of the Evolved 3GPP System

We elaborate here how the "internals" of the Evolved 3GPP system work. We stick to the most natural order, so that it becomes visible how single, smaller steps realize these main procedures:

- access control (authentication and authorization),
- node selection sunctions,
- IP address allocation and configuration,
- initial attachment and detachment,
- intra and inter system mobility (idle and active mode),
- session handling,
- specialized procedures (ANDSF communication, service continuity, CS fallback, warning message delivery), and
- procedures for I-WLAN mobility.

Note that this chapter provides explanations on functional level. The names of procedures and messages presented here may deviate from the actual names of protocol messages.

5.1 Access Control (Authentication and Authorization)

This function has already been discussed within the security concept in sub-section 4.8 and is included as one block in initial attachment (see sub-section 5.4) and Tracking Area Update (see sub-section 5.8.1). Here we present the corresponding information flows.

5.1.1 Authentication and Authorization in E-UTRAN access

Figure 5-1 shows several message exchanges related to authentication and key agreement for a UE in E-UTRAN:

- a procedure for determining the identity of a UE,
- authentication data download from HSS/Authentication Center to MME,
- challenge of the UE,
- context transfer.

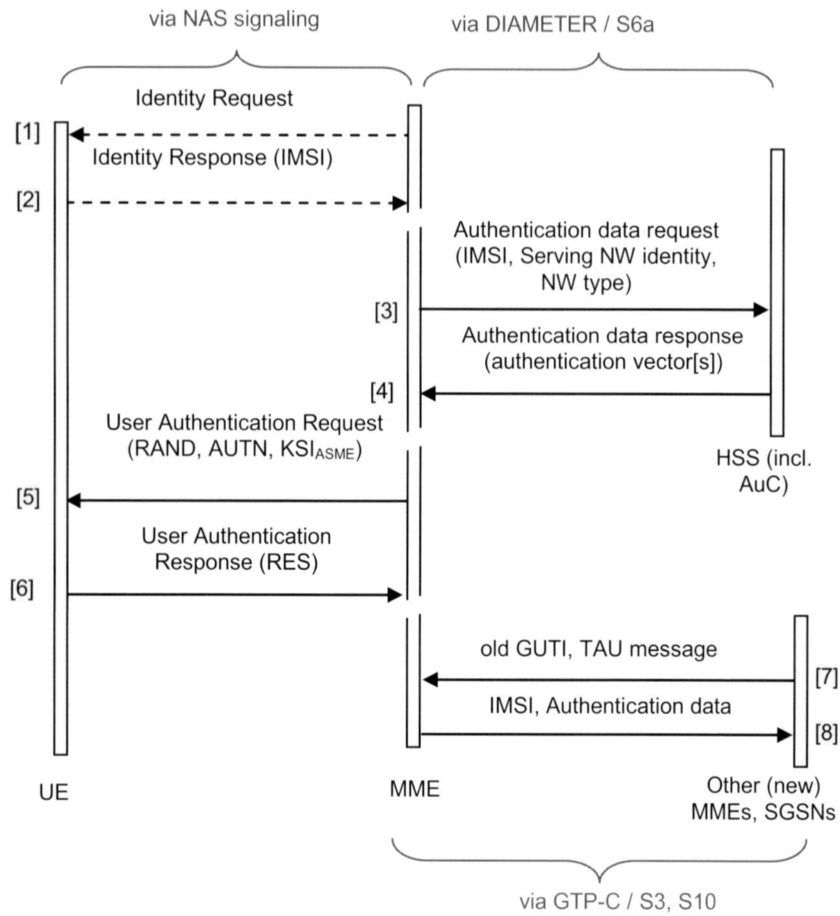

Figure 5-1: message exchanges related to authentication and key agreement in E-UTRAN

The steps are:

1. If the (temporary or permanent) identity of the UE is not known/derivable by the MME, it can at any time request an identification from the UE by issuing the corresponding NAS signaling message (see sub-section 6.9).
2. The UE responds with its permanent identity (IMSI).

3. Before MME can request the UE for authentication/authorization it needs to have available an authentication vector. In contrast to the signaling in the legacy case, there is no real need that the MME fetches more than one EPS authentication vector at a time from HSS/AuC (although it is possible); the more elaborate key hierarchy reduces the need to perform AKA runs. The request and response (in step 4) are mapped onto DIAMETER commands Authentication-Information-Request/Answer (AIR/AIA), see sub-section 6.5.
4. The authentication vectors themselves are derived from the legacy authentication vectors by using a different Key Derivation Function. They consist of of a quadruple: random number (RAND), expected result of authentication (XRES), authentication token (AUTN) and the main key for access security (KASME).
5. The MME sends the AUTN and K_{ASME} to the UE, together with the challenge (random number).
6. The UE answers the challenge by sending the result of the security hash function (RES).
7. At some later point in time, as required by inter system change, another control plane node (now new MME or SGSN), asks for security context transfer; this transfer cannot be done unprotected, the requesting node therefore has to deliver the whole NAS signaling message (in this example a TAU), together with the temporary identification of the UE.
8. The (now old) MME checks the NAS signaling message with the current security context, and if it is found valid, transfers it to the new requesting node.

5.1.2 Authentication and Authorization in non-3GPP access

5.1.2.1 Trusted non-3GPP access

The overall message flow given in Figure 5-2 splits into (1) the EAP [1] front-end part between UE and EAP authenticator in the trusted non-3GPP access network, (2) the EAP relay transfer via DIAMETER signaling between trusted non-3GPP access network and AAA server, and (3) the authentication vector plus user profile download and final registration between AAA server and HSS. In case of roaming an additional AAA proxy would appear between the untrusted non-3GPP access network and the AAA server. This procedure builds heavily on the one already used for I-WLAN in the legacy system, the main differences being:

- the use of a different key derivation function,
- more/different amount of information transferred in the DIAMETER messages, and
- indication for trust relationship and IP mobility mode selection within EAP payload.

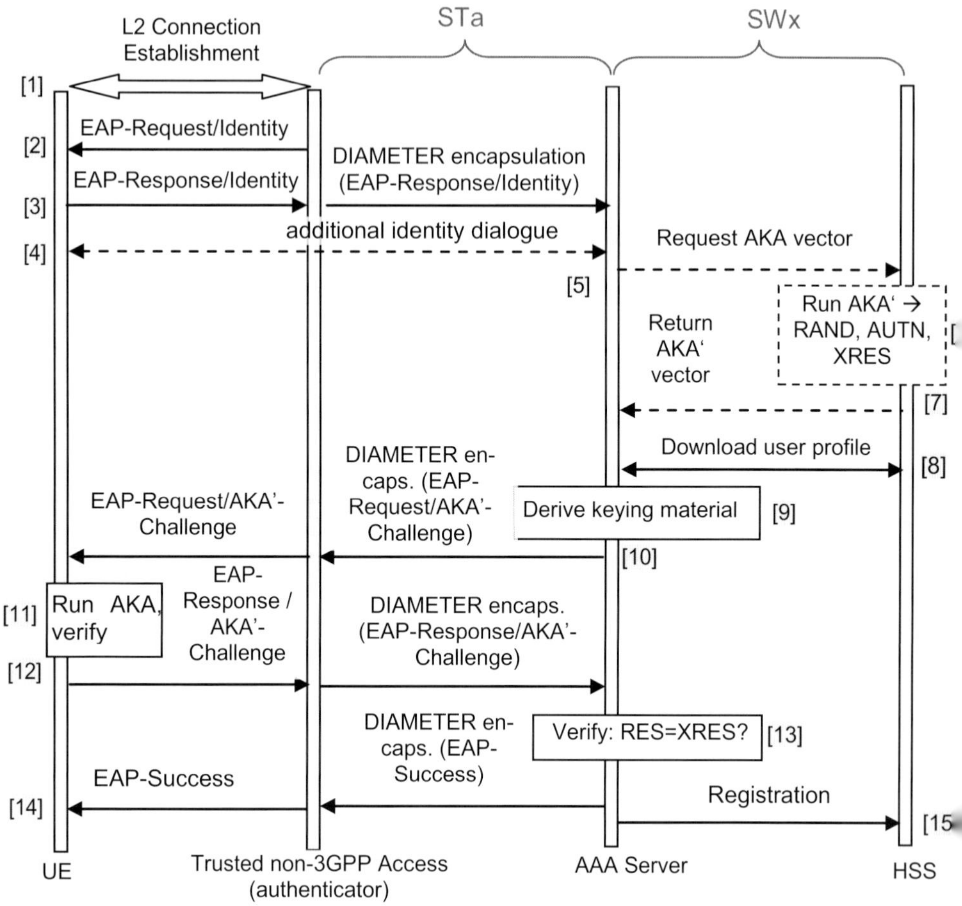

Figure 5-2: message exchange for authentication and key agreement in trusted non-3GPP access

In a simplified manner, the signaling (across the L2 link in the non-3GPP access network as well as over the backend AAA interfaces STa and SWx) and the node internal procedures are:

1. The UE establishes L2 connectivity in the trusted non-3GPP access; this step is specific to any access technology.
2. The EAP-authenticator in the trusted non-3GPP access issues an identity request; the underlying transport for EAP is out of scope here and may vary per access technology.

3. The UE sends an EAP response to the authenticator in the trusted non-3GPP access and uses the NAI format of its identity (for details of other parameters see sub-section 6.3). UE's response is encapsulated as EAP-payload in a DIAMETER request towards the AAA server, together with the access network identity.
4. An additional identity dialogue might be needed (in the case that intermediate nodes in the AAA infrastructure have changed the identity provided by the UE; this could/should be avoided, but in general it is not excluded.)
5. If no remaining, unused authentication vector is stored on the AAA server, it is fetched from HSS/AuC. Authentication vectors depend on the access network identity, so if the UE moves between access networks, authentication vectors downloaded earlier (when the old access network identity was presented) are no longer valid. Furthermore, as AAA server and HSS reside both in the home network, there is no big advantage in fetching more than one vector anyway.
6. The HSS/AuC chooses a random number and runs EAP-AKA' (**prime**, with a different key derivation function than 3GPP's original EAP-AKA), as this is identified from the access network identity as the required authentication algorithm. The result is XRES (expected result) for the challenge towards UE, AUTN (Authentication Token) for authentication of the network towards the UE, C_K' (ciphering key) and I_K'(integrity key).
7. The HSS returns the authentication vector to the AAA server. If the HSS has already assigned an AAA server for the UE, it refers the requesting AAA server to that one.
8. If it is not yet available, the AAA server downloads the authorization profile for the UE from the HSS.
9. The AAA server derives the EAP keying material MSK (Master Session Key) and EMSK (Extended Master Session Key) from C_K' and I_K'. This is according to the IETF RFC [2] , which was newly created for 3GPP's needs in this context. At this point a new pseudonym or fast re-authentication identity can be chosen (see below); if so, it will be ciphered and integrity protected with this keying material.
10. The EAP-Request with the AKA' challenge for the UE is encapsulated in DIAMETER and sent to the authenticator in the trusted non-3GPP access network, optionally with new pseudonym and fast re-authentication identities, and these are stored. The request is de-capsulated from DIAMETER message and forwarded on EAP transport level to the UE.
11. The UE runs the EAP-AKA' procedure locally and checks AUTN; if successful, the keying material is calculated. If pseudonym and/or re-authentication id were received, they are stored by the UE.
12. The UE returns the answer to the challenge to the authenticator in the trusted non-3GPP access network, ciphered and integrity protected by the received keys. The authenticator encapsulates the EAP-response RES for the AKA' challenge in DIAMETER message and forwards it to the AAA server.

13. The AAA server compares RES with the expected result XRES.
14. If the check is successful, the EAP success message is sent via the authenticator to the UE. In the DIAMETER message, additional keying material can be included for local use by the authenticator.
15. The AAA server registers the UE in HSS.

It is not always necessary to run a full authentication. Depending on operator's policy, a fast re-authentication scheme may be used, which allows to reduce the amount of signaling. The message flow can thus be compacted, but then the security level is also slightly decreased (because the presentation of the original credentials is not necessary). We refer to specialized sources for details [2].

5.1.2.2 Untrusted non-3GPP access

For untrusted non-3GPP access, it is optional to run a 3GPP based access authentication, i.e. between an authenticator in the access network and the AAA server. If it is used, the message flow given in Figure 5-2 applies, but with some reduced scope of parameters (see sub-section 6.5).

The mandatory IPSec tunnel authentication and authorization proceeds as shown in Figure 5-3, for a full tunnel authentication and authorization. It is based again on EAP-AKA (not EAP-AKA prime, though), but this time the EAP payload is encapsulated in IKEv2 signaling between UE and ePDG.

ePDG acts as EAP authenticator in this case.

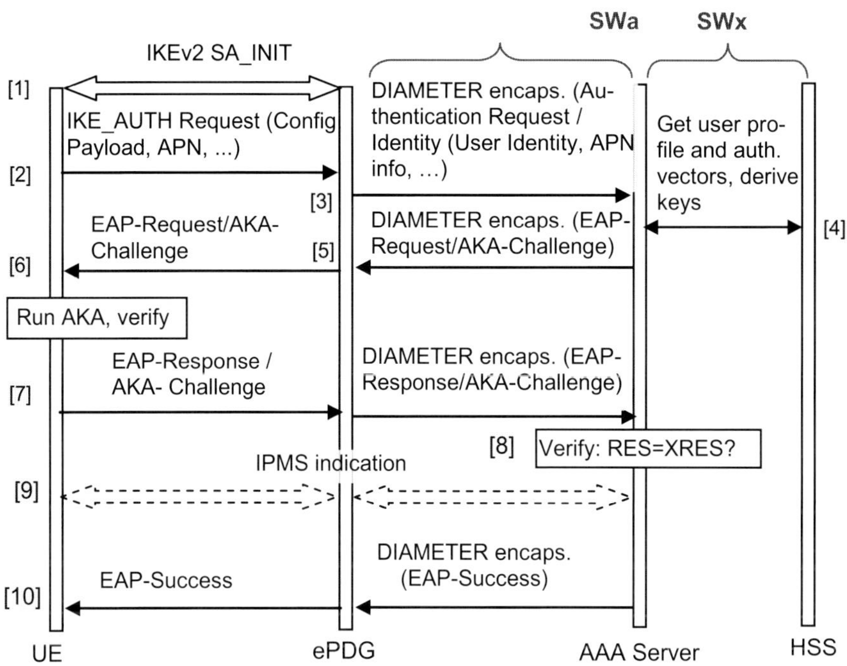

Figure 5-3: message exchange for authentication and key agreement in untrusted non-3GPP access

As a prerequisite, the UE has attached on L2 to the non-3GPP access and configured a local IP address (this part is not shown here), and selected the ePDG (see 5.2.5). The subsequent steps are:

1. The UE initiates IKEv2 signaling with ePDG (SA_INIT dialogue); a negotiation about cryptographic algorithms is done and a primary security association is established.
2. UE initiates the IKE_AUTH request, carrying EAP various payloads (identification of UE by NAI, requested IP address information, APN); yet, the AUTH parameter itself is left out, indicating that EAP over IKEv2 is to be used. The UE may also indicate that it supports the MOBIKE [3] extension of IKEv2, in which case later handovers to other untrusted non-3GPP access networks can be optimized (see sub-sections 5.10.3 and 6.4). A child security association is then established. The requested IP address information can be either the IPv4 address, IPv6 home address prefix or Home Agent address.

3. The ePDG extracts the EAP payload from IKEv2 and encapsulates it in a DIAMETER message to the AAA server, presenting the identity of the UE and also including APN information. The realm part of the NAI (see sub-section 4.21.3) indicates that authentication is requested for a UE wishing to access the EPC via ePDG, and **not** via a legacy PDG (the network needs to be able to distinguish the case, due to the different service properties involved).
4. If it has no more unused authentication vectors available, the AAA server fetches them from HSS, together with the user profile (static QoS data); for that purpose the IMSI has to be extracted from the NAI, or deduced from the pseudonym. It is checked that the subscription allows the pending setup of the IPSec tunnel, i.e. access from an untrusted non-3GPP network and below the number limits for tunnels (if the limit is reached, the oldest security association, i.e. IPSec tunnel, is requested to be deleted by the respective ePDG).
5. The AAA server encapsulates the challenge for the EAP-AKA request in a DIAMETER message and sends to the ePDG.
6. The ePDG, in turn, extracts the EAP payload from DIAMETER and places it into a payload for the next part of the IKEv2 AUTH dialogue, including also authentication data for itself. This completes the child security association setup with the UE. The UE checks the authentication parameters and runs the AKA procedure.
7. The UE responds to the ePDG with the result of the EAP-AKA challenge (as EAP payload); from ePDG it is forwarded to the AAA server within a DIAMETER message.
8. The AAA server verifies that the result of the challenge received from UE matches the expected result (as known from the currently used authentication vector).
9. If dynamic indication of IP mobility mode is employed, the mobility mode is selected and indicated to the UE in an additional EAP-AKA notification dialogue between the AAA server and the UE.
10. The success of the authentication is notified to the UE, via the ePDG.

Again, similar to the trusted non-3GPP access authentication, a "fast" variant of tunnel authentication and authorization is also defined.

5.2 Node Selection Functions

5.2.1 General

Several nodes need to be determined initially, when the UE attaches for the first time, or subsequently in course of idle mode mobility, active mode mobility (handover) and also for load balancing reasons. The detailed selection procedure themselves are not standardized, they typically would take into account the network structure/configuration and load. Standardized are only the general sequence and input/output from nodes. Logically the selection is done within the chain eNodeB → MME → Serving GW → PDN GW.

The selection of the eNodeB is done by the UE, in a process known as "camping on a cell" (and the cell belongs to one eNodeB); radio conditions like signal strength are used to find a suitable cell (see sub-section 4.2.5.4).

With further enhancements of the architecture, e.g. for Local IP access (see sub-section 3.10.2), it is likely that more intelligence has to be put into the node selection. As an example, if a PDN GW close to the radio access (or co-located with the eNodeB) is selected, for the benefit of offloading local, home network or Internet destined traffic efficiently, and the UE moves on, this "anchoring" becomes unfavourable at some point. It may then be necessary to do a an optimized PDN GW re-allocation in idle mode.

5.2.2 Selection of MME

The eNodeB selects the MME for a UE in proportion to a weight factor per MME, which is received from MMEs with connectivity to one eNodeB via S1-AP messages. The weight factor represents the capacity of an MME relative to other MMEs. For load distribution reasons a MME may release S1-connections with an appropriate cause (preferably in UEs idle mode); in this case the eNodeB will bar the MME for re-selection.

Also EPC nodes, namely source MME in case of MME relocation (e.g. for the purpose of load balancing) and SGSN in case of intersystem mobility, may need to select an MME.

5.2.3 Selection of Serving GW

The MME determines a Serving GW by location and load (on potential Serving GWs). Additional criteria are the EPC protocol variants to be supported (GTP and/or PMIP), this variation is per PLMN. MME needs therefore corresponding configuration data on PLMN granularity (which would reflect the status of roaming agreements). Also information on presence of combined nodes (Serving GW and PDN GW co-located) is to be taken into account for the selection of Serving GW.

The basic mechanism for determining the concrete IP address of the Serving GW, while additionally applying above criteria, is by DNS (resolving a characteristic FQDN).

5.2.4 Selection of PDN GW

In 3GPP access, the PDN GW selection is also performed by MME, based on subscription data in HSS and APN input from the UE; the procedure is modeled after GGSN selection in GPRS. Either the identity of a PDN GW and APN, or an APN and the indication whether a PDN GW from VPLMN or HPLMN is to be selected is received during the HSS interrogation e.g. during the initial attach procedure. If a handover from non-3GPP access occurs, the PDN GW is already known and a new PDN GW selection is avoided. If the identity of the PDN GW is provided in the form of an IP address, no further resolution is necessary; if it is a FQDN, an additional step of DNS query needs to be performed. In this way the network operator gains flexibility in traffic distribution. The procedure is given schematically in Figure 5-4 for a roaming case.

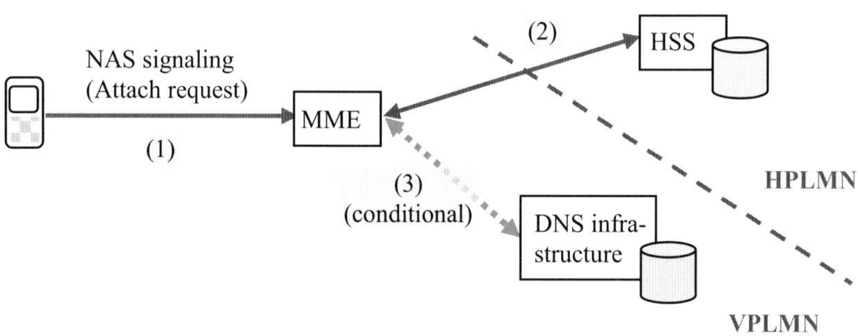

Figure 5-4: PDN GW selection (schematically)

For connectivity to an additional PDN a further PDN GW selection is done, based on the APN given by the UE.

The PDN GW selection for UEs in untrusted non-3GPP access involves no MME; instead it is done instead by the AAA server during the authentication / authorization procedure.

5.2.5 Selection of ePDG

As a most simple but rather inflexible option, the ePDG address may be statically configured in the UE. Additionally, a a more flexible dynamic discovery is possible and probably preferred in larger networks and subscriber base. It is based on a DNS query. First the UE has to connect on L2, in an access specific way, to the untrusted non-3GPP access network and configure its IP layer. The UE then builds a FQDN specific for ePDG nodes and issues a DNS request.

Depending on the capabilities of the access specific signaling, a list of PLMN identifications may have been provided to the UE (this happens e.g. in the PLMN selection according to legacy I-WLAN procedures). In this case the UE can include such a received PLMN identification in the FQDN for ePDG (in which case it aims at finding an ePDG in a VPLMN); else, the UE includes the HPLMN identification in the FQDN. The overall procedure is visualized in Figure 5-5 for the latter case, under the assumption that the local DNS infrastructure in the untrusted nin-3GPP access network forwards the DNS request to the HPLMN for resolution of the FQDN to an IP address.

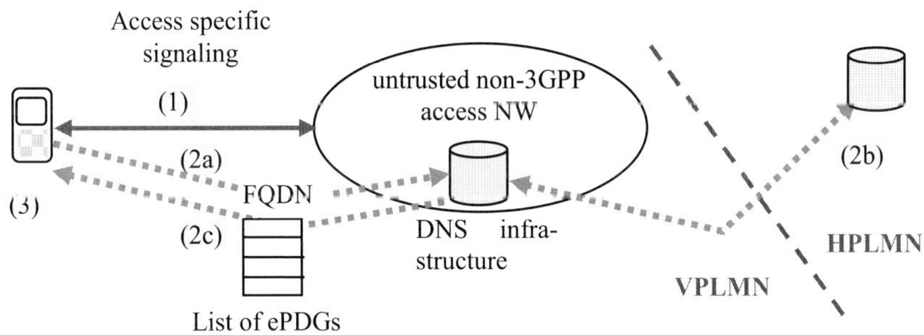

Figure 5-5: ePDG selection (schematically)

If the ePDG is to be selected in course of a handover and there was none allocated before, the UE includes the identification of the previously used PLMN in the FQDN for the DNS query.

The result of the query may be not only one, but a list of IP addresses for ePDGs, and they can also belong to different IP version; the UE then selects one out of the list, with the same IP version as its local IP address.

In case of connectivity with multiple PDNs (e.g. for accessing a corporate network and the 3GPP operator's services at the same time) also only one ePDG would be selected (this corresponds to the restriction for only one Serving GW in 3GPP access). But one very special exception exists: in course of handovers between different non-3GPP access technologies (both of which would need to be considered untrusted), where the same ePDG is not accessible from the target non-3GPP access, the UE may start the IKE signaling and IPSec tunnel establishment in parallel; this is intended for reducing latency in signaling and thus more performant handover.

5.2.6 Selection of Home Agent (for DSMIPv6)

A Home Agent is to be discovered and selected before the DSMIPv6 mobility scheme can be employed (see also sub-section 4.15.3); this is the case for DSMIPv6 use in the integrated SAE/EPC as well as in the standalone configuration for I-WLAN mobility as described in sub-section 4.16. The possibilities are manifold:

1) Static pre-configuration: the UE is provisioned with operator's settings.
2) DNS query: the UE builds a well-known FQDN for a Home Agent and issues the request to the DNS, which resolves it to one or more IP addresses.
3) DHCPv6 signaling: the UE actively requests in broadcast mode (from the nearest DHCP local server) or unicast (if the dedicated DHCP server is known) the Home Agent as an IP configuration parameter.
4) Protocol configuration options in NAS signaling (if the UE is in 3GPP access) or PMIPv6 signaling (if the UE is in trusted non-3GPP access).
5) IKEv2 signaling during IPSec tunnel setup with ePDG (applies only if UE makes access via an untrusted non-3GPP access network).

5.3 IP Address Allocation and Configuration

5.3.1 Basic concepts

In a packet switched domain like the Evolved 3GPP system, applications use IP for transferring data. A general prerequisite is therefore to allocate one or more IP addresses to the UE; in 3GPP access this is tightly related to the EPS bearer concept (see sub-section 4.6), meaning that an EPS bearer provides all means to transport IP packets.

The matter is complicated by the need to care for different IP address versions: a UE may want to have, and a PDN may allocate, either an IPv4 or an IPv6 address. But also the "dual stack" case exists, where IPv4 and IPv6 addresses may be used simultaneously. Also, the concept of how IP addresses are allocated already differs from the base protocols in use.

We need to discuss here five issues:

- when is an IP address is allocated/configured,
- type of IP address (IPv4/IPv6) requested/supported by the UE and granted/supported by the network,
- multiplicity regarding protocol stack (single or dual stack) and connectivity (i.e. multiple PDN feature, see sub-section 4.20),
- access variant (3GPP or non-3GPP access), and
- signaling protocol variant (GTP or PMIPv6 in 3GPP access; PMIPv6, DSMIPv6 or MIPv4 in non-3GPP access).

Originally the intention was to integrate IP address allocation fully with default bearer establishment within the PDN connection establishment (in particular with the PDN CONNECTIVITY REQUEST, either as a standalone request or within initial ATTACH REQUEST, see sub-sections 4.9 and 4.10), so that an "always-on" behavior is achieved. This would be in contrast to GPRS, where an IP address is allocated only for PDP contexts, and these may be established and torn down on demand within the attached state. However, the need for support of non-integrated UEs (i.e. those where the L2 and L3 protocol stack are implemented separately, typically laptop computers with a LTE radio interface card) led to the weakening of this requirement. These devices may first establish a default bearer without an IP address, and subsequently perform the necessary configuration by the DHCP protocol (as an IP "bootstrapping" mechanism).

Table 5-1 lists the different cases for IP address handling:

<table>
<tr><th>UE wants IP@ type …</th><th>At default bearer setup (NAS signaling) UE …</th><th></th><th>After default bearer setup UE …</th><th>Later UE may …</th></tr>
<tr><td rowspan="4">IPv4 only</td><td>sets PDN type to IPv4</td><td></td><td>----</td><td>configure additional parameters via DHCPv4</td></tr>
<tr><td colspan="4">OR</td></tr>
<tr><td>sets PDN type to IPv4</td><td rowspan="2">AND</td><td rowspan="2">requests IPv4 address via DHCPv4 signaling</td><td>request renewal and release IPv4 @</td></tr>
<tr><td>sets flag in PCO for IP@ allocation via DHCPv4</td><td>UE may configure additional parameters</td></tr>
<tr><td>IPv6 only</td><td>sets PDN type to IPv6</td><td>AND</td><td>does IPv6@ auto-configuration</td><td>----</td></tr>
<tr><td>IPv4 and IPv6</td><td>sets PDN type to IPv4v6</td><td></td><td></td><td></td></tr>
</table>

Table 5-1: IP address handling scenarios / part 1

There may be restrictions with regards to the dual stack capability or preference in the network and subscription. Also, as a matter of fact, the legacy 3GPP system supports only single address bearers. The dependencies are described in Table 5-2.

It includes also the repetition case, where the UE requested originally dual stack connectivity (i.e. IPv4v6 address allocation); it then received the indication by a cause #52 that the network restricts it to single IP version and subsequently the UE requests additional PDN connectivity to the same PDN with the other IP version.

UE capability	**Request PDN type (for this APN)**	**Restriction *)**	**Result**	**Comment**
IPv4	IPv4	none	IPv4	
		IPv6	Reject	
IPv6	IPv6	none	IPv6	
		IPv4	Reject	
unknown	IPv4v6		same as for UE capability IPv4v6	
IPv4v6	If no IP@ assigned yet	IPv4v6	IPv4	Cause #50; no further attempt for IPv6 allowed
			IPv6	Cause #51; no further attempt for IPv4 allowed
			Single (IPv4 or IPv6)	Cause #52; no further attempt for IPv4v6 allowed
	If IPv4@ assigned, cause #52 already received.	IPv6	**)	
	If IPv6@ assigned, cause #52 already received.	IPv4	**)	

*) restriction may be subscription, PDN GW capability or operator preference **) repetition case, restrictions are already considered

Table 5-2: IP address handling scenarios / part 2

The row "unknown" in the table reflects the case of non-integrated UEs, where there is a separation between MT (Mobile Termination, which is the endpoint for the NAS signaling) and TE (Terminal Equipment).

5.3.2 IPv4 address allocation and configuration via DHCPv4

In this case the "dummy" IPv4 address 0.0.0.0 is allocated during default bearer establishment. The PDN GW acts as the DHCP server for the subsequent request, according to Figure 5-6; the IPv4 address needs to be configured for an external PDN, PDN GW can act as a DHCPv4 client towards the DHCP server in this external network. There is a difference between PMIP and GTP based EPC: in the former case the Serving GW acts additionally as a DHCP relay agent. Note that from EPC point of view all DHCP signaling happens already in the user plane. Apart from initial request for IPv4 address allocation (which is valid only for a certain period of time), a message exchange for lease time extension eventually becomes necessary.

Parameters configurable via DHCPv4 are available for L2, L3 and L4, e.g. default TTL (time-to-live), MTU (maximum transmission unit), default router flag, TTL for TCP etc. (see IETF RFC 2131 [4]).

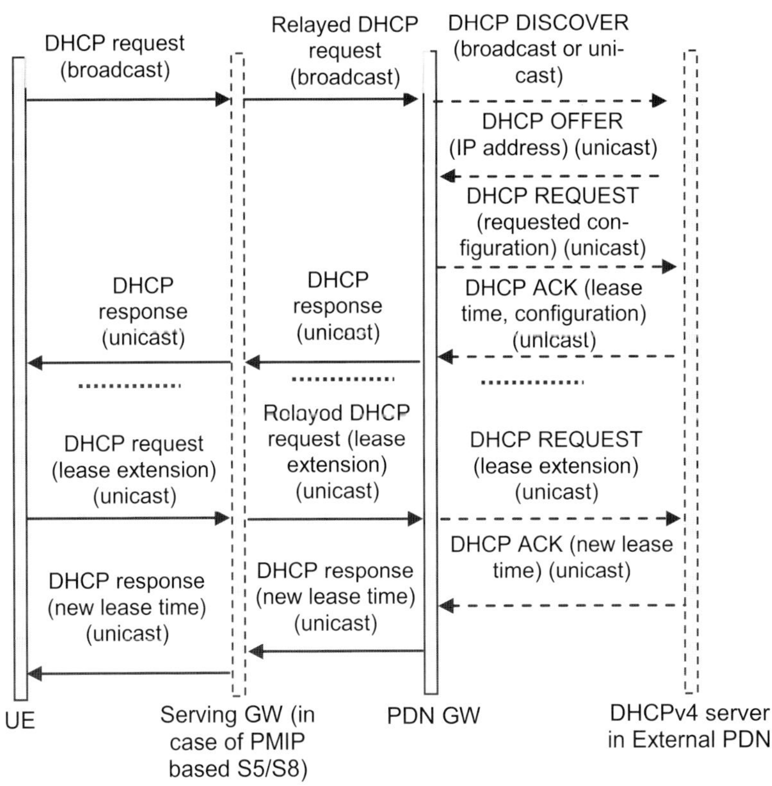

Figure 5-6: IPv4 address allocation and parameter configuration via DHCPv4

5.3.3 IPv6 address autoconfiguration

If the UE wants to configure an IPv6 address, this should happen according to the standardized mechanism in IPv6 architecture, i.e. based on the advertisement of an IPv6 address prefix by the access router and completion by the UE, generating the remaining part of the IP address. It is shown in Figure 5-7, for both GTP and PMIP based S5/S8. The characteristic difference is that the access router function is performed by the PDN GW in the former case and by Serving GW in the latter.

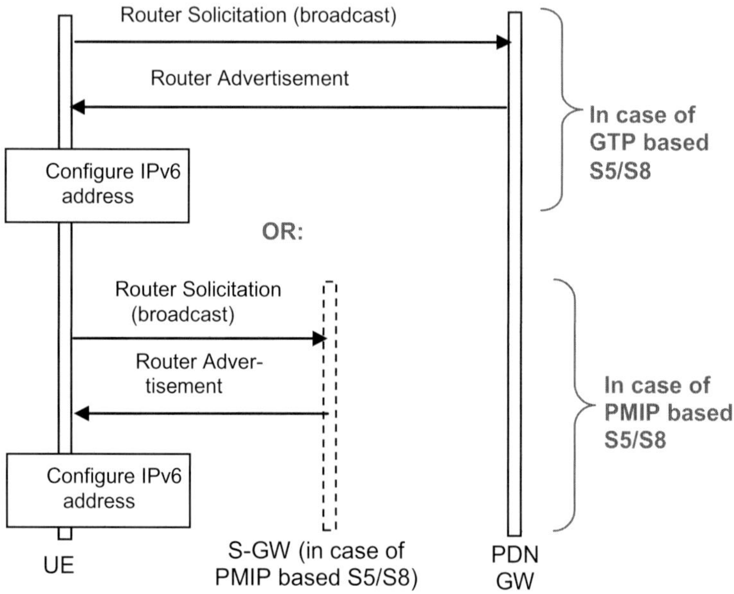

Figure 5-7: IPv6 address autoconfiguration

5.3.4 IP address configuration via DHCPv6

It is an option to retrieve further IPv6 related configuration data from a server via DHCPv6 after IPv6 address auto-configuration. A message flow similar to (IPv4) address configuration and parameter configuration via DHCPv4 applies, but is not shown here. Again, the DHCP server may reside on PDN GW or in an external PDN. The difference with respect to GTP and PMIP based S5/S8 is again that in the latter case Serving GW is involved as a DHCPv6 relay.

5.4 Initial Attachment

The purpose of the attach procedure is identify the UE against the network, bootstrap a secure context shared between UE and network, and to establish the default bearer.

5.4.1 Initial Attachment in E-UTRAN

The procedure applies at switch on within the E-UTRAN coverage and is realized by the message flow sketched in a slightly simplified form in Figure 5-8 for the case of the S5/S8 protocol being GTP (the variant for PMIP is presented further below).

Some variation is found when the UE has still an old context assigned (from a previous attachment); in this case the "old" MME would be interrogated at step 3, and step 9 would include the removal of subscription data in the old MME (triggered from the HSS) and potentially tear-down of any old resources with GWs (Serving and PDN GWs). This part is not shown in the graph. A similar procedure applies also if the attachment occurs in course of a handover.

The flow describes the successful case; many negative cases are possible and indicated by corresponding causes in the ATTACH REJECT message, like illegal/unkown UE, EPS service / PLMN / TA / cell / CSG not allowed. The UE can react upon that (e.g. search for another cell) and the attachment can be repeatedly tried, up to 5 times (an attach attempt counter is incremented for every unsuccessful case). Because the session handling is piggybacked onto the mobility handling in the attach procedure, also a negative outcome of the ESM part leads to a failure of the EMM procedure.

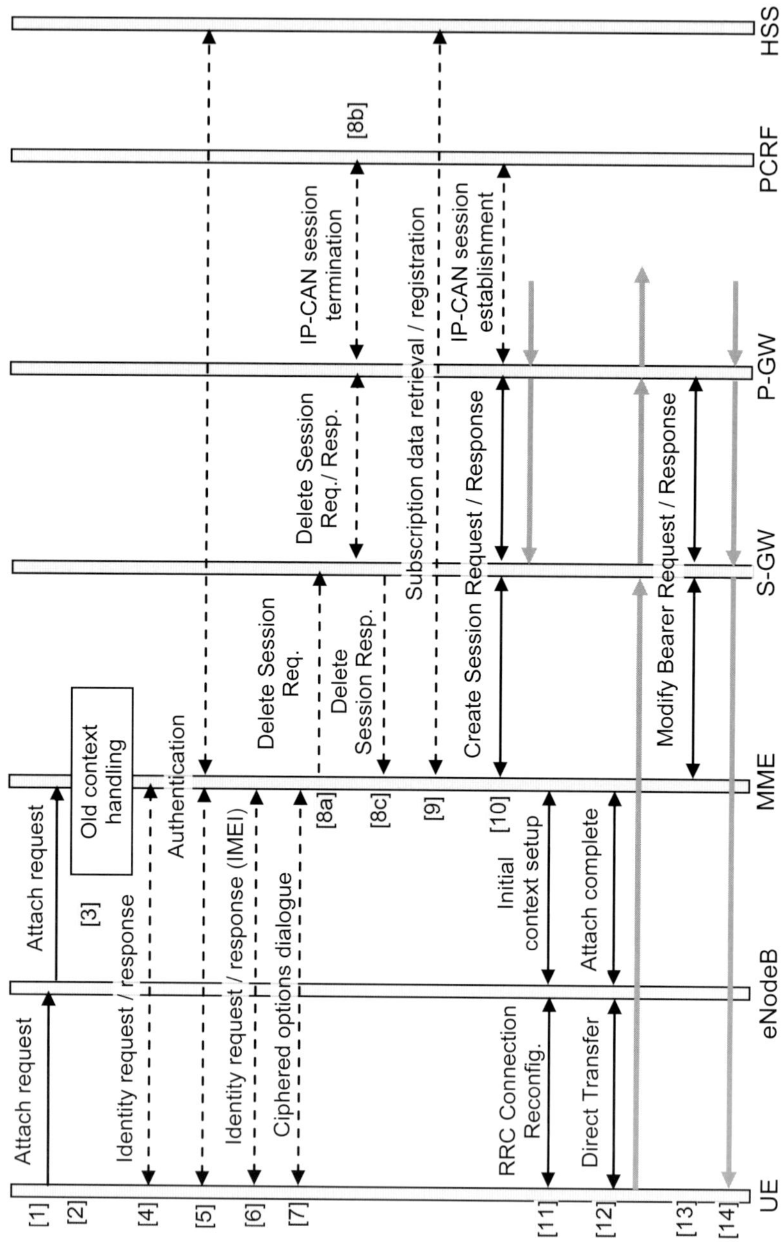

Figure 5-8: initial attachment in E-UTRAN (GTP used on S5/S8)

We see this quite elaborate message flow:

1. The UE sends an attach request to the eNodeB with (among other parameters) its own identity, optionally old context related identities, identity of the selected network (for the case that there it is a shared network deployment), UE's capabilities, type of the attach and type of the requested PDN, security context (if already existing), optionally requested APN and protocol configuration options (PCO), if their ciphering is not required; alternatively, the indication for ciphering of APN and PCO would be included instead.
2. The eNodeB derives an MME identification, if possible, from the old context and the selected network. Alternatively the eNodeB selects by itself an MME (this is also done if the eNodeB has no connection with the derived MME). Then the eNodeB sends the attach request to this MME.
3. If the MME is able (from old context related identities) to derive the identification of an SGSN or MME, where the UE was handled before the last detach, it retrieves the old context data from there (this step is not shown in the graph, for simplicity); retrieval of the context from a GERAN/UTRAN node is security protected (by P-TMSI signature). If the MME is the same as before the last detach, and an EPS context is still available, it will be used by the MME.
4. If the MME cannot find an existing context for the UE, it sends an identification requests to the UE. The UE responds with its IMSI.
5. The authentication procedure may be performed between the network and the UE; it is mandatory, if no context was found for the UE, or the attach request sent by the UE was not integrity protected, or if integrity check failed. Otherwise it is optional. This step also (re)initializes the NAS security setup, and subsequent NAS signaling messages are protected from now on.
6. If not yet done in a combined manner in step 5, the MME may retrieve the equipment identity (IMEI) of the UE and subsequently check it by interrogating EIR (this latter interrogation is not show here, for simplicity).
7. If the UE indicated in step 1 that PCOs and/or APN need to be transferred ciphered, they shall now be retrieved by the MME.
8. If old resources are present (e.g. to incomplete previous teardown with detach), the MME triggers signaling towards Serving GW for clearance (Delete Session Request/Response), and this leads to corresponding signaling onwards to PDN GW; if dynamic PCC is deployed, the IP-CAN session with PCRF is also terminated.
9. The MME retrieves UE's subscription data from the HSS (Update Location) and updates the registration status there. If the HSS has still stored the registration of the UE with the previous SGSN or MME, it initiates deletion of data on these nodes; this will further lead to clearance of still allocated resources in previously used Serving GW/PDN GW(s) and potentially PCRF, similar to step 8. These latter two steps are not shown in the graph, for simplicity.

10. MME requests the setup of bearer(s) from Serving GW; Serving GW executes signaling for establishment of the user plane with PDN GW. If dynamic PCC is employed, the PDN GW interacts with PCRF for the establishment or, if this is an attachment with handover indication, modification of the IP-CAN session. The PCRF responds with the now applicable PCC rules. Alternatively the PDN GW can use local static policies. The PDN GW allocates IP address information. The result of this step are established bearers between Serving GW and PDN GW and buffering of data in Serving GW (i.e. the first downlink IP packets may arrive).
11. The MME exchanges with eNodeB a signaling dialogue for initial context setup. The eNodeB performs RRC connection reconfiguration dialogue with the UE. In the message to the UE the 'attach accept' indication of the NAS signaling layer (between MME and UE) is included.
12. The UE sends the direct transfer message on the S1-MME interface to the MME, including the 'attach complete' indication of the NAS signaling layer (between UE and MME). The first uplink data may now be sent.
13. The MME signals to Serving GW the modification of bearers; this is forwarded to the PDN GW and acknowledged back.
14. From now on the Serving GW forwards packets (including the buffered ones) to the UE via the eNodeB.

If the S5/S8 protocol is PMIP, we find some characteristic differences with PCRF interaction and with tunnel handling between Serving GW and PDN GW. Note, however, that all steps involving the PCRF occur only if dynamic PCC is deployed; if this is not the case, the PDN GW can use static policies.

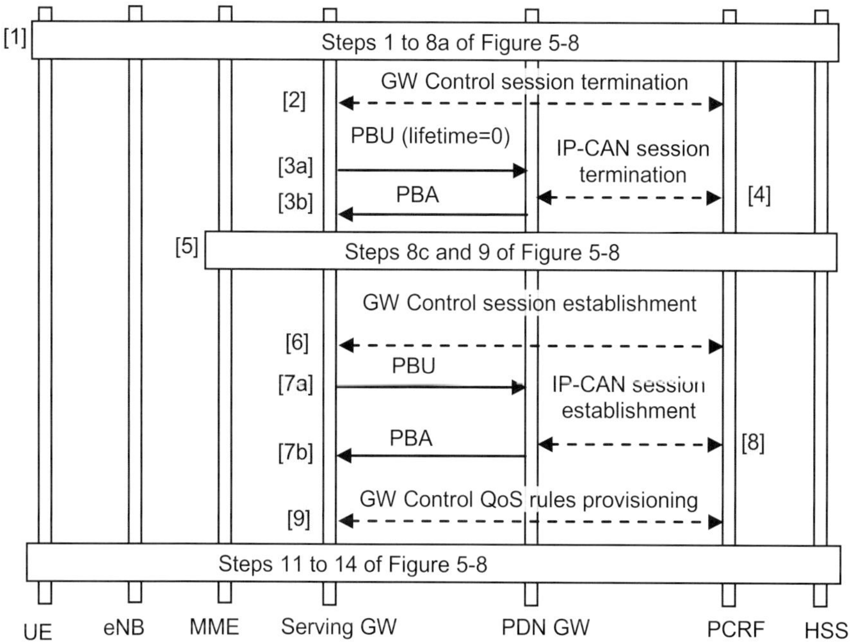

Figure 5-9: initial attachment in E-UTRAN (PMIP used on S5/S8)

1. The initial steps of attachment are performed as in the case with GTP as EPC mobility protocol variant (node selection, request for attachment, old context handling, identity handling, authentication, ciphered options handling)
2. Serving GW (actually the BBERF contained in it) terminates the GW control session with the PCRF.
3. The MAG function in Serving GW exchanges PMIP signaling with the LMA in PDN GW by sending a Proxy Binding Update (PBU) message with lifetime zero; this erases the old Binding Cache entry in the LMA. The acknowledgement is by a Proxy Binding Acknowledge (PBA).
4. The PDN GW tears down the old IP-CAN session with PCRF.
5. The teardown of old sessions is communicated back to the MME, and the signaling for subscription data download and registration with HSS is performed like in the flows for GTP (this is independent of the EPC protocol variant).
6. The Serving GW/BBERF then establishes the (new) GW control session with PCRF.
7. The MAG function in Serving GW exchanges the PMIP signaling with PDN GW, creating also the mobility tunnel between them.
8. The PDN GW establishes the IP-CAN session for dynamic PCC (if deployed).

9. The PCRF also informs the Service GW of the QoS rules to be applied in the E-UTRAN access; they are used in the next step to install radio bearers accordingly.
10. The remaining parts of the initial attachment occur as for the GTP based variant of EPC protocol.

5.4.2 Initial Attachment in legacy 3GPP PS access

The initial attach to the EPC from GERAN/UTRAN access occurs with a PDP context activation; a prerequisite is that the SGSN is Rel. 8 capable. The difference with the normal case (i.e. PDP context activation via Gn/Gp) lies in the messagies for session creation and modification of bearers (after radio bearers have been established) by the SGSN; normally these are exchanged with a GGSN, now they need to be addressed to a Serving GW; from there they are relayed to a PDN GW. These differing steps are shown in Figure 5-10 for the case that GTP is used on the S5/S8 interface.

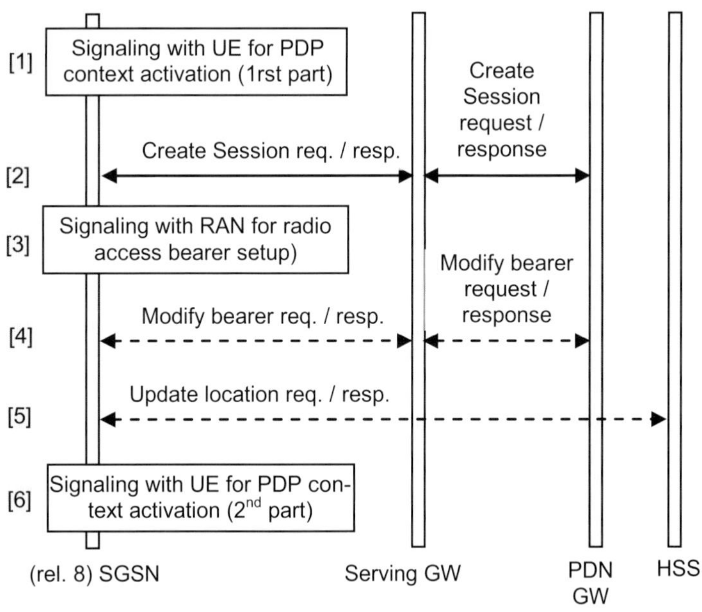

Figure 5-10: initial attachment in legacy 3GPP PS access

The procedure consists of building blocks in the legacy PS system and signalling with and within EPC:

1. The UE requests a PDP context activation via NAS signaling from SGSN.
2. The SGSN selects both a PDN GW and a Serving GW (corresponding to the similar function in MME). Create Session Request / Response message pairs are exchanged between SGSN and Serving GW, and between Serving GW and PDN GW. Among many other parameters, UE identification, PDN GW address, APN (if requested by the UE), PDN type (that is, requested IP version) and RAT type are conveyed to the Serving GW, and the latter two parameters also to the PDN GW. These are GTP-Cv2 control messages. Return parameters are, among others, tunnel parameters, the IP version selected by the network, APN (confirmed or newly allocated) and IP address information.
3. The SGSN performs signaling with the GERAN/UTRAN, to let establish the radio access bearers and thus reserve resources on the radio link.
4. Signaling for modification of the bearers along SGSN – Serving GW – PDN GW is executed; if the resources on the radio access could not fulfill the requests, the procedure is terminated. This signaling is also used to establish a direct tunnel between RNC and Serving GW, if this is supported in the network and by the configuration. Also, the signaling is used to start the flow of packets, if the PDP context activation occurred in course of a handover with non-3GPP access; before, buffering of packets would be done at the Serving GW.
5. If dynamic policies are employed in the network, the PDN GW would interact with the PCRF in the manner shown as in the initial attach procedure in E-UTRAN (see previous sub-section).

5.4.3 Initial Attachment in Trusted non-3GPP access

Although one could imagine distributed arrangements, it is assumed for simplicity that all interworking functions in the non-3GPP access are performed by one node, which is called Gateway (GW) here. The GW also plays the role of an EAP authenticator. Figure 5-11 depicts the procedure.

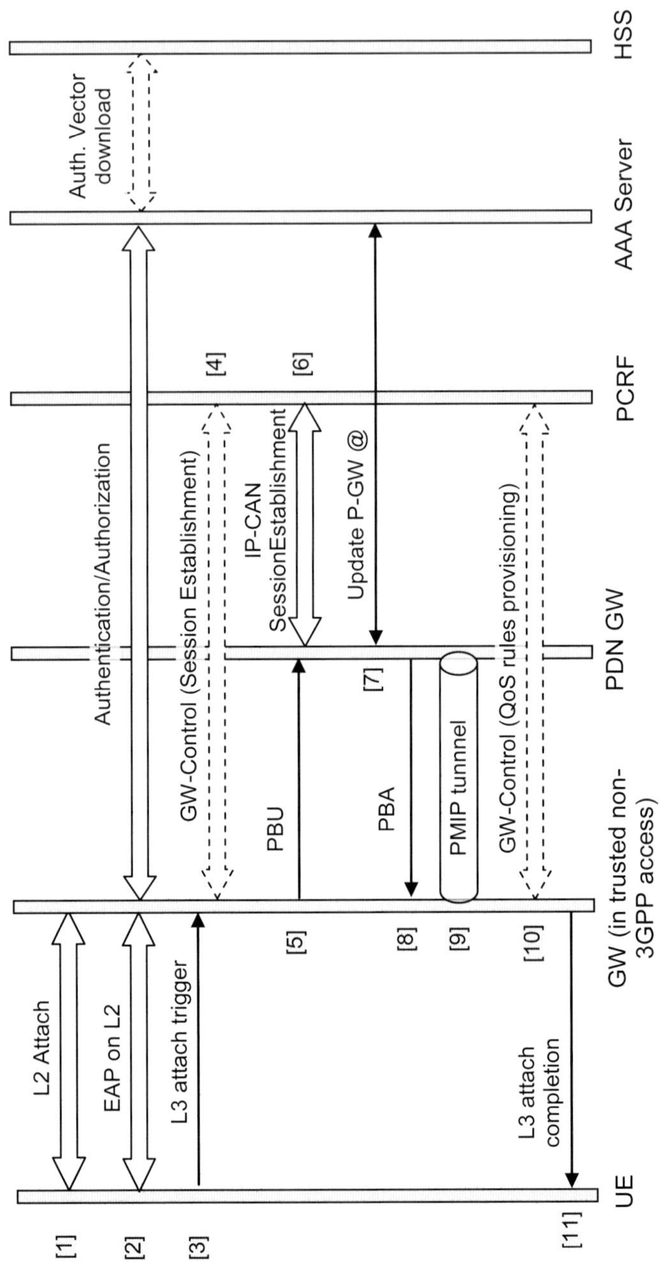

Figure 5-11: initial attachment in trusted non-3GPP access (with PMIPv6)

The steps are the following:

1. The UE attaches on L2 to the trusted non-3GPP access network and is handled on a GW there. This procedure is access specific.
2. The EAP dialogue on L2 is triggered by the GW. After the response has been received from the UE, it is encapsulated in a DIAMETER authentication/authorization request message and sent towards the AAA server. The EAP-AKA' (**prime**) method is executed with the UE. HSS is interrogated only if no unused authentication vector was left on the AAA server.
3. A message triggering the L3 attach is sent from the UE to the GW; it is access specific. If supported by the access technology, an APN provided by the UE could be conveyed.
4. Optionally (if dynamic QoS provisioning is deployed for the trusted non-3GPP access), the GW requests establishment of the control session with the PCRF.
5. The GW sends a Proxy Binding Update message to the PDN GW, including the NAI of the UE, the requested IP address information (which kind of IP address, IPv4 and/or IPv6), the APN (necessary, as the PDN GW may connect to several PDNs; it may be the APN requested by the UE or assigned during authentication/authorization procedure in step 3).
6. The PDN GW requests establishment of the IP-CAN session from the PCRF.
7. The AAA server is updated with the PDN GW address.
8. The PDN GW sends a Proxy Binding Acknowledge message back to the GW, including the assigned IP address information (IPv6 prefix and/or IPv4 address).
9. The PMIPv6 tunnel between the GW and the PDN GW is now ready for data transfer.
10. Optionally (if dynamic QoS provisioning is in place for the trusted non-3GPP access), the PCRF can update the QoS rules.
11. An indication for L3 completion is sent from the GW to the UE (it is access specific). The assigned IP address information is conveyed to the UE, together with the finally selected APN and protocol configuration options (if these are supported in the trusted non-3GPP access). At this point full IP connectivity is achieved.

5.4.4 Initial attachment in untrusted non-3GPP access

A precondition is that the UE has acquired a local IP address from the untrusted non-3GPP access network. The overall steps to be performed are:

- ePDG discovery/selection (see sub-section 5.2.5),
- IKEv2 signaling and IPSec tunnel establishment,
- PDN GW selection (see sub-section 4.2.3) and
- PMIPv6 tunnel establishment.

The IPSec tunnel acts as a virtual L2 connection for the UE and the ePDG takes the role of the access router, as seen from higher protocol layers. Two cases of mobility management are supported, network based mobility with PMIPv6 and client based mobility with DSMIPv6. In the former case, the ePDG acts as MAG and the PDN GW as an LMA, in the latter case the PDN GW plays the role of a Home Agent.

The detailed message flow is shown in Figure 5-12. In a roaming situation (i.e. when the PDN GW is in a VPLMN) an AAA proxy would be located between ePDG and the AAA server. The procedure applies if IPMS had determined that S2b (that is PMIPv6) is to be used; note that even in this case the HA address for use with DSMIPv6 (e.g. at a later time) may be requested by the UE.

Before the attachment flow below is executed, an additional authentication could be run; it is not important for the security of the EPC (and therefore not specified as mandatory), but it could be used to bootstrap L2 security.

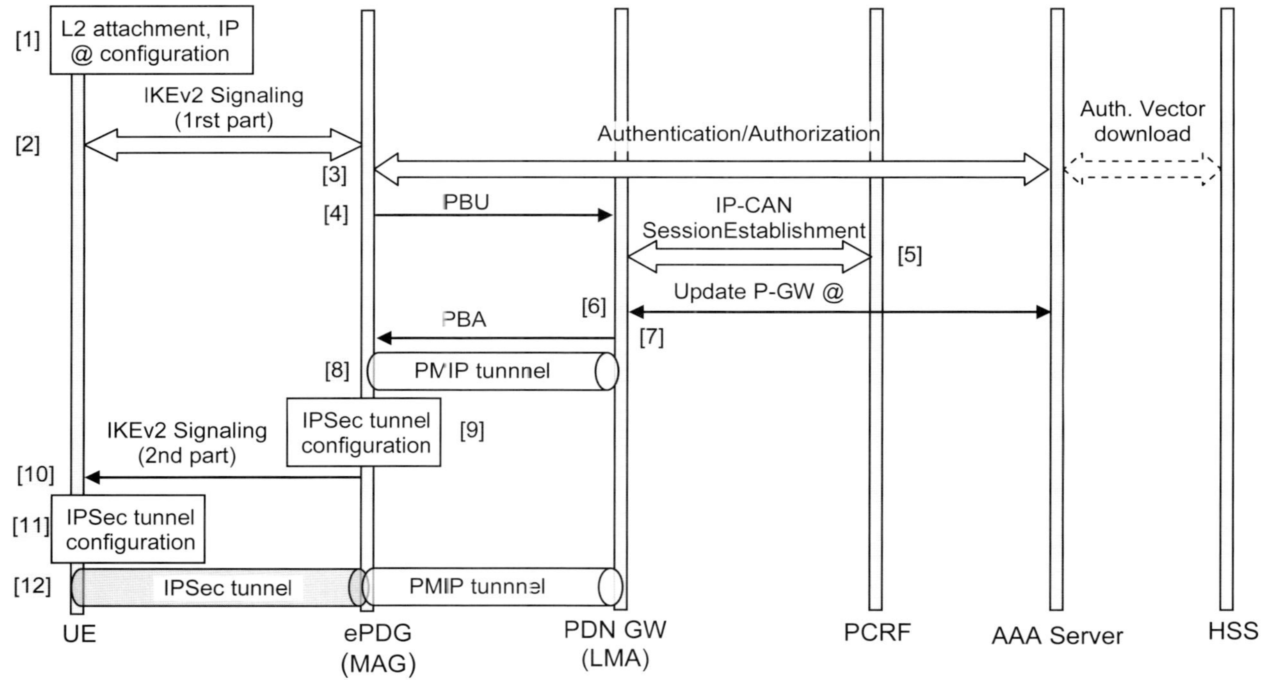

Figure 5-12: initial attachment in untrusted non-3GPP access (with PMIPv6)

1. The UE has attached on L2 to the untrusted non-3GPP access network and received a local IP address.
2. The initial part of the IKEv2 signaling is performed (INIT dialogue and UE's AUTH request). In the AUTH payloads conveyed to the network the UE requests IP address information, optionally support for MOBIKE and optionally the request for HA address (if DSMIPv6 is intended to be used e.g. later). The APN may be included in this request (if not anyway the default APN/service is to be used).
3. The authentication and authorization dialogue across the SWm interface is performed between ePDG and the AAA server (see details in sub-section 5.1.2.2). The interaction with the HSS is necessary, if no unused authentication vector is available in the AAA server.
4. The ePDG sends a Proxy Binding Update message to the PDN GW, including the NAI of the UE, the requested IP address information (i.e. which kind of IP address, IPv4 and/or IPv6 is requested), the APN (necessary, as the PDN GW may connect to several PDNs; it may be the APN requested by the UE or assigned during authentication/authorization procedure in step 3).
5. The PDN GW requests establishment of the IP-CAN session from the PCRF.
6. The AAA server is updated with the PDN GW address.
7. The PDN GW sends a Proxy Binding Acknowledge message back to the ePDG, including the assigned IP address information (IPv6 prefix and/or IPv4 address).
8. The PMIPv6 tunnel between the ePDG and the PDN GW is now ready.
9. The IPSec tunnel configuration is installed at the ePDG, together with the mapping of tunnels.
10. The IKEv2 dialogue is completed by sending back the response part of the AUTH dialogue. It includes configuration payloads carrying the IP address information and, if it was requested by the UE, also the HA address.
11. The IPSec tunnel configuration is installed on the UE.
12. Now the IPsec tunnel between UE and ePDG, and the PMIPv6 tunnel between ePDG and PDN GW are concatenated and user traffic can flow.

The information flow for attachment during handover is similar, with these differences: the UE would already have assigned IP address information (IPv6 prefix and/or IPv4 address). This is indicated also to the AAA server, so that no new PDN GW selection is performed.

5.5 UE requested (additional) PDN connectivity

The precondition for this procedure is that a (default) PDN connection already exists; in this case the UE may at any time request connectivity to another PDN, by indicating another APN. This may also occur in course of a handover. PDN GW selection is performed by the MME and may either lead to the selection of the same or another PDN GW (in case of handover of course the already allocated PDN GW is used). For the IP address allocation the same rules apply as for the initial attachment. The additional PDN connection will be provided with a default bearer and potentially, depending on the resource requirements, one or more dedicated bearers. Figure 5-13 shows the information flow, with GTP as the EPC protocol.

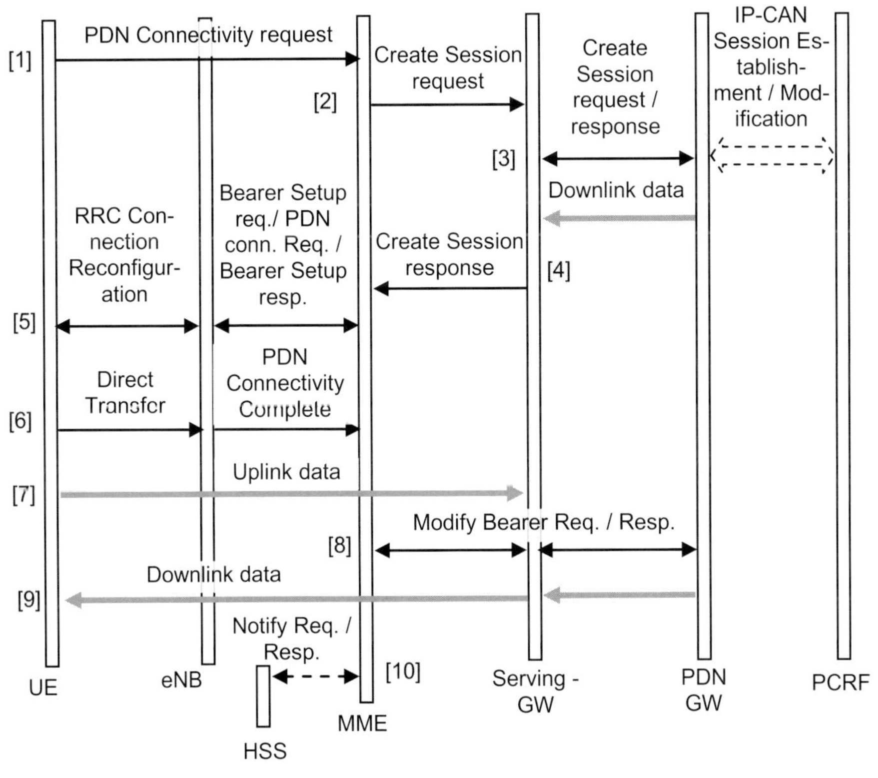

Figure 5-13: UE requested additional PDN connectivity

The procedure starts from the UE and proceeds with preparation of resources in the core network, followed by the establishment of radio resources, as given by the following steps:

1. The UE sends the NAS signaling message "PDN Connectivity Request" to the MME, indicating APN, type of requested IP address and whether it is within a handover.
2. MME checks if the APN is authorized; if so, it requests the establishment of a bearer session from Serving GW and supplies the necessary QoS information.
3. The Serving GW exchanges signaling for bearer session establishment with the PDN GW. If dynamic policies are employed, the PDN GW also informs the PCRF by performing a IP-CAN session Modification or Establishment procedure (depending if it is within a handover or not). The PCRF provide PCC rules and may modifiy QoS parameters (which may lead to the creation of dedicated bearers). The PDN GW provides session related information back to the Serving GW. At this point also downlink data can be forwarded from the PDN GW to the Serving GW, if this is not a handover case (in case of handover, routing of data packets from the PDN GW to the Serving GW starts only after step 8).
4. The Serving GW confirms the preparation of resources in the core network to the MME and provides the relevant data (in particular allocated IP address and QoS data).
5. The MME instructs the eNodeB to proced with the setup of (radio access) bearers. eNodeB performs RRC Connection Reconfiguration with the UE to put in place the appropriate resources on the radio link. The corresponding signaling messages are acknowledged back. Depending on the radio resource status, QoS could be modified.
6. The UE sends back a PDN Connectivity Response message to the MME (which is transported in a S1-AP Direct Transfer message first to the eNodeB).
7. From now on uplink data packets can be sent by the UE.
8. The MME exchanges with the Serving GW update signaling (Modify Bearer procedure).
9. If this is a handover case, data packets can now also be forwarded by the PDN GW to the Serving GW. The Serving GW can forward packets to the UE as soon as it received the response message from PDN GW in step 8.
10. The MME informs the HSS, if necessary (e.g. if it is not a handover case, a different PDN GW was chosen, etc.).

The equivalent procedures for a UE in non-3GPP access (trusted via PMIPv6, DSMIPv6 or MIPv4 FA mobility protocols, and untrusted) are not presented here; they have larger parts in common with the corresponding initial attachments.

5.6 Detachment

Various reasons lead to a detachment: switch off of the UE, unsuccessful re-authentication, resource limitations, withdrawal/change of subscription etc. Thus the detachment can be initiated by the UE or by by various network nodes (MME, SGSN or HSS).

5.6.1 UE initiated Detach from E-UTRAN

Figure 5-14 presents the message flow for UE initiated detach from E-UTRAN for the protocol GTP on S5/S8.

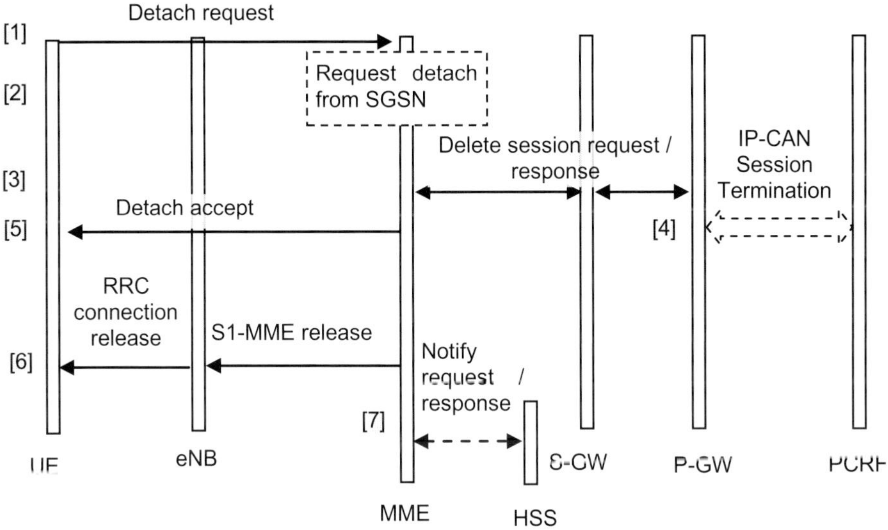

Figure 5-14: UE initiated detach from E-UTRAN

1. The UE is attached to E-UTRAN and requests detachment by sending the NAS signaling message “Detach Request” to the MME. If the detachment is for switch off, the UE would not wait for the outcome of the procedure, i.e. for step 5.
2. Optionally, if ISR is activated, the MME informs the associated SGSN and requests it to clear resources between the SGSN and the Serving GW (not shown in the graph).

3. Between MME, Serving GW and PDN GW the "Delete Session request/response" signaling message pairs are exchanged. If user location information was requested for this case from PDN GW, the Serving GW includes the most recent one (from either MME or SGSN of step 2).
4. If dynamic PCC is employed in the network, PDN GW runs the IP-CAN session termination procedure with PCRF.
5. If the detach was not for switch off, the MME sends back a "Detach Accept" message to the UE via NAS signaling.
6. The UE performs the S1-MME release with the eNodeB. This leads subsequently to the release of the RRC connection between eNodeB and UE and release of the radio access bearers between eNodeB and Serving GW.
7. If the subscription data indicates that the UE is allowed to handover to a non-3GPP access, and depending on MMEs configuration, the MME can notify the HSS about the detachment.

5.6.2 Detach from Trusted non-3GPP Access

Figure 5-15 visualizes the procedure for detach in trusted non-3GPP access when PMIPv6 is used (non-roaming; in case of roaming a AAA proxy and a V-PCRF in the VPLMN would additionally be present in the signaling paths). The internal structure of the trusted non-3GPP access is not in scope of 3GPP, there may be a co-location or distribution of functionality; for simplicity, we assume that the interfaces towards the EPC (PDN GW and PCRF) are handled by one node (access GW).

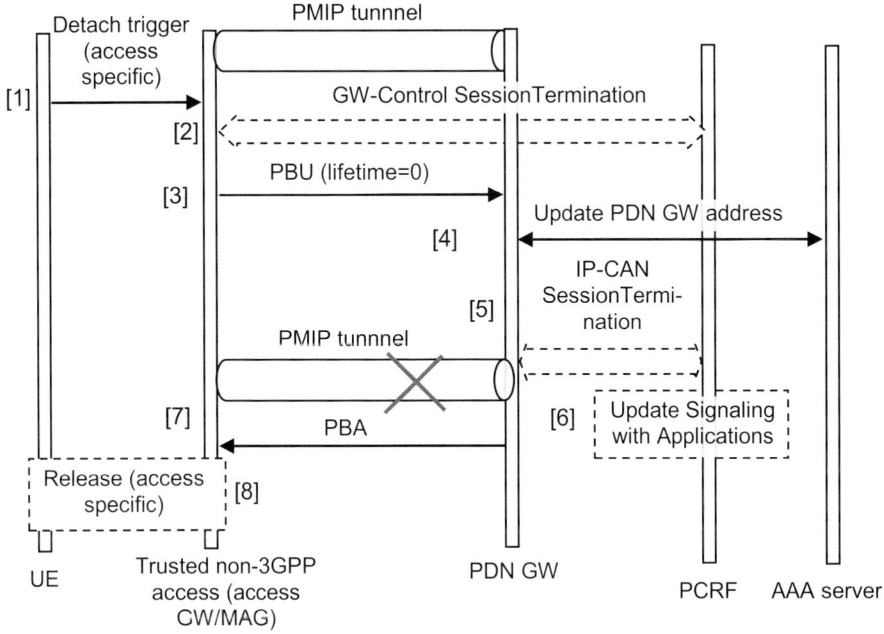

Figure 5-15: detach from Trusted non-3GPP access

1. The UE initiates a detach in the trusted non-3GPP access, in an access specific way.
2. The access GW in trusted non-3GPP access network initates a GW-Control Session Termination (exchanging a DIAMETER Credit-Control Request /Response message pair with the PCRF).
3. The PMIPv6 node (MAG e.g. in the access GW) in trusted non-3GPP access network sends a Proxy Binding Update with lifetime zero to the PDN GW (LMA in PMIP terms), including also the identification of the UE and the APN;
4. The PDN GW deletes its registration in the AAA server.
5. The PDN GW terminates the IP-CAN session with the PCRF, exchanging a DIAMETER Credit-Control Request/Response message pair.
6. If applications are bound to the IP-CAN session (e.g. due to SIP/SDP via P-CSCF), the PCRF performs corresponding update signaling.
7. The PDN GW deletes the binding cache entry and sends back a Proxy Binding Acknowledgement message back to the MAG in the trusted non-3GPP access network, indicating the identification of the UE and the APN. The PMIP tunnels is thus deleted.

8. An access specific release procedure between the trusted non-3GPP access and the UW may be performed (this is out of 3GPP's scope).

If the UE was attached to multiple PDNs, steps 2 to 7 have to be repeated for each PDN connection.

5.7 S1 connection management

5.7.1 Service Request

This procedure is used to bring the UE from idle into active mode and thus establishes the Radio Bearers. Two sub-variants exist, one triggered by the UE and the other one triggered by the network.

The message flow for the UE triggered Service Request (for GTP as EPC protocol) is seen in Figure 5-16.

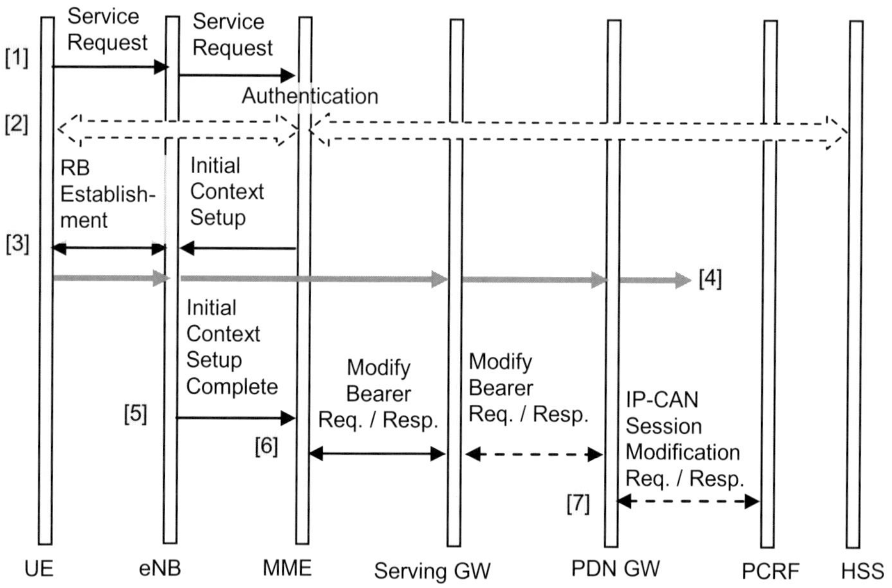

Figure 5-16: UE triggered Service Request procedure (GTP based S5/S8)

The procedure comprises the following steps:

1. The UE in idle mode needs to send uplink signaling or data; it issues a Service Request message by NAS signaling towards the eNodeB, which forwards it to the MME.
2. The network optionally executes a (re-)authentication procedure, involving the UE and HSS.
3. After successful authentication, the MME instructs the eNodeB to establish the UE context (Initial Context Setup via S1AP); this leads to establishment of Radio Bearers.
4. At this point uplink data can start to flow from the UE (via eNodeB, Serving GW and PDN GW).
5. The eNodeB acknowledges the establishment of the UE context back to MME (S1AP message).
6. The MME requests the modification (activation) of bearers from Serving GW. If the RAT type has changed, the Serving GW requests also to update the bearers for the S5/S8 connection from PDN GW.
7. If the RAT type has changed, and if dynamic PCC is deployed, this event has to be propagted to the PCRF, in order to download updated policy rules. If dynamic PCC is not deployed, the PDN GW can use statically provisioned rules.

If the UE is in idle mode and the network needs to perform signaling or deliver user data (e.g. terminating traffic arrives), it uses the network triggered Service Request procedure (Figure 5-17).

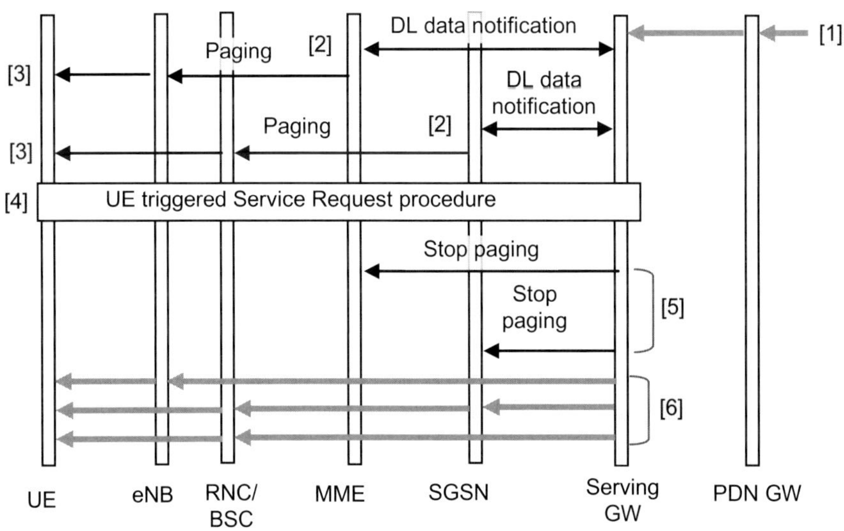

Figure 5-17: network triggered Service Request procedure

The necessary steps here are:

1. Downlink data arrives at the PDN GW and is forwarded to the Serving GW; if the UE is not in connected mode, this data is buffered.
2. A corresponding "downlink data notification" message is sent to MME and/or SGSN, depending on to which node(s) control plane connectivity is established (i.e. on the ISR status).
3. The MME and/or SGSN (depending on at which nodes the UE is registered) send paging messages to the respective RAN nodes, from where the paging is forwarded over the radio channels.
4. As paging response, the UE performs the UE triggered Service request procedure either in E-UTRAN (as described in the previous sub-section), or the corresponding procedure in GERAN/UTRAN access.
5. If no response is received within a configurable time, the paging can be repeated (but the details of repetition handling are not standardized). If ISR is activated the UE can respond from either one of the two accesses; if a paging response is received from one access, a message is sent from the Serving GW to stop paging in the other access.
6. From now on downlink data is forwarded from the Serving GW, via the respective access, to the UE. Note that in case of use of Direct Tunnel (see also subsection 3.4.2) with UTRAN, the data does not pass the SGSN GW, but is sent directly from Serving GW to the RNC.

5.7.2 S1 Release

This procedure releases the logical connection (on the S1AP interface) between eNodeB and MME; it leads also to the teardown of S1 bearer resources between eNodeB and Serving GW, as well as deletion of the UE's context in eNodeB. The UE is moved from active/connected to idle state. The reason and trigger for S1 release may come from eNodeB (e.g. user inactivity, UE request for signaling conenction release, failures at eNodeB), or from MME (e.g. authentication failure).

The resulting message flow is quite simple, see Figure 5-18.

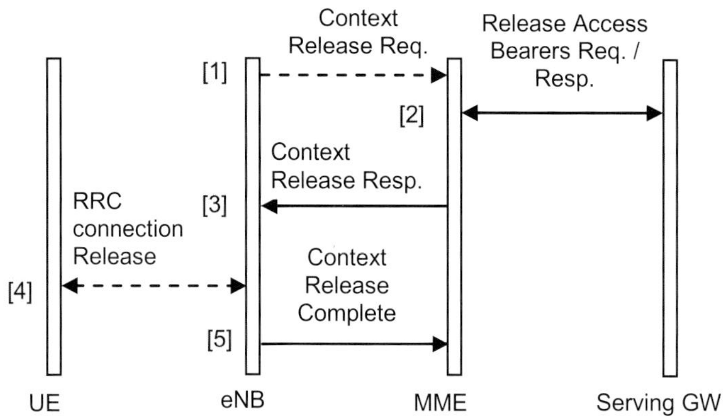

Figure 5-18: S1 release procedure

1. Only in case a trigger for S1 connection release is detected: the eNodeB sends the triggering S1AP message for release to the MME, indicating the appropriate cause (examples are given above).
2. The MME instructs the Serving GW (either caused by step 1, or by an own trigger) to teardown the resources of all S1-U (i.e. radio access) bearers. If the corresponding reporting was requested previously by the PDN GW, the MME includes UE's location. The Serving GW releases all UE context information related to the eNodeB; it prepares itself for buffering and performing network triggered Service Request procedure, if subsequently downlink data packets arrive.
3. The MME confirms the release back to the eNodeB.
4. If the RRC has not yet been released (intermediately), it is now down by signaling to the UE; the eNodeB releases its UE context.
5. The eNodeB confirms the release back to the MME.

5.8 Intra-System Mobility

These functions comprise idle mode mobility (Tracking Area Update) and active mode mobility (i.e. handovers in several forms).

5.8.1 Tracking Area Update (TAU)

TAU is a procedure initiated by the UE to update the registration status with the network. There are quite many reasons for the UE doing a TAU, and it can be in idle or active mode:

- when the UE moves into a new Tracking Area (TA) (see sub-section 4.5);
- periodically, based on a timer (independent of the TA);
- inter-system change (from legacy system to the so-called S1 mode, from HRPD to S1 mode);
- for MME load balancing, triggered by a RRC release with corresponding cause;
- to inform the network if certain UE specific paramaters have changed;
- for recovery from some error cases;
- when the UE comes back to E-UTRAN after CS fallback;
- after manual selection of a CSG cell whose identity is not in UE's Allowed CSG list.

Variations of the TAU procedure are possible depending on whether the Serving GW and/or MME is changed or not; Figure 5-19 presents the message flow for TAU with Serving GW change and MME change, and for GTP as the S5/S8 protocol. The UE can include a parameter "last visited Tracking Area", which can be used by the MME to assign TAIs to the UE in an optimal way; e.g. if it is noticed that the UE traverses a metro area from east to west, the MME could assign one more western TA, so that the UE can move on longer in idle mode without signaling to the network.

The procedure consists of:

1. Following one of the above listed triggers, the UE sends a TAU request to the eNodeB, including its (old) temporary identity, identification of the last visited TA, security parameters and bearer status, potentially indicating the selected network (in case of network sharing), old MME (if existing) and the need for load balancing.
2. The eNodeB a determines the MME, including load conditions and presence of old MME identification; it then sends the TAU request to the MME, adding the identity and TAI of UE's current cell.
3. From the old, global temporary identification the MME may derive the old MME or SGSN, and fetches the old context of the UE; in this step security protection is maintained; because the new MME does not yet have UE's context (in particular no security context), it sends the whole TAU request message for checking with the old security context to the old MME/SGSN. By this step also the security context in the new MME is initialized.

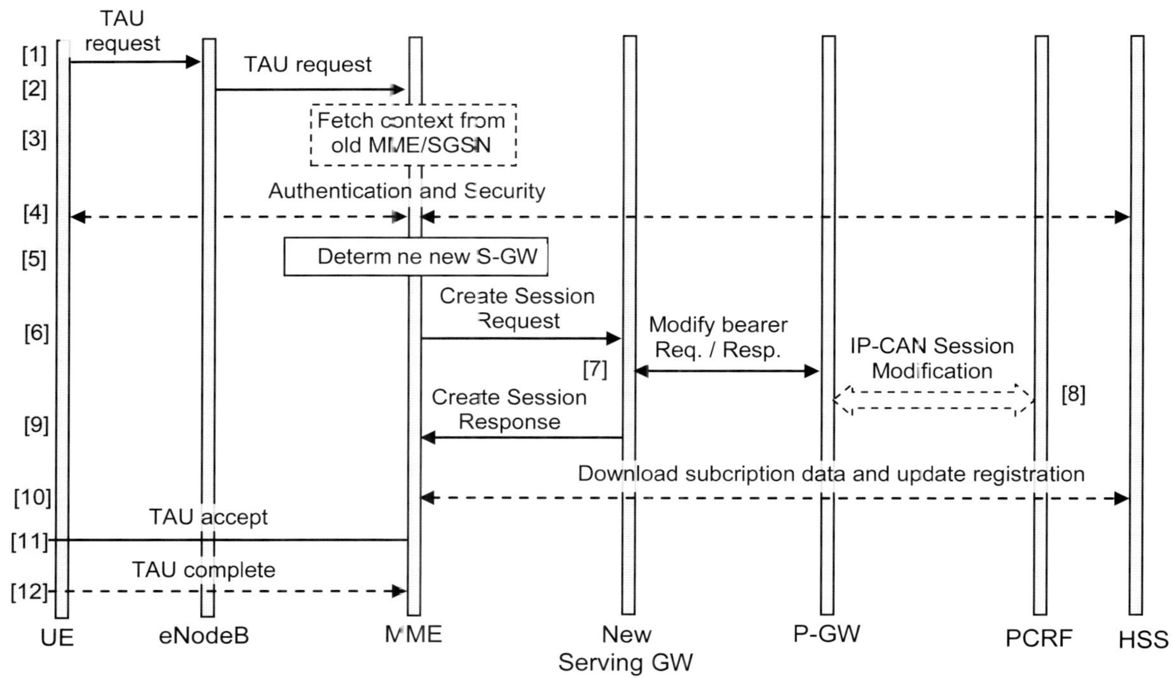

Figure 5-19: Tracking Area Update (in E-UTRAN)

4. If no old context was available, or the security check on the old context fails, an authentication must be performed and involves MME, UE and HSS.
5. The new MME determines that a new Serving GW has to be used (e.g. due to load reasons or user plane path optimization). At this point the new MME also informs the old MME/SGSN about this fact (not shown in the graph).
6. The new MME creates a new context for the UE; the bearer status received from UE is cross checked with the information received from the old MME/SGSN and synchronized by releasing old resources. If no old bearer context exists, the TAU is rejected.The MME sends a "Create Session" message to the new Serving GW.
7. The Serving GW exchanges a message pair for modification of the bearers, so that they are installed between itself and the PDN GW.
8. If dynamic PCC is employed, the PDN GW performs signaling for IP-CAN session modification with the PCRF. If this procedure happens in course of an intersystem change, at this point the change of RAT would be reported.
9. The Serving GW confirms the modification of bearers with the PDN GW back to the MME.
10. The MME retrieves UE's subscription data from the HSS (Update Location) and updates the registration status there. If the HSS has still stored the registration of the UE with the previous SGSN or MME, it initiates the deletion of data on these nodes; this will further lead to clearance of still allocated resources in the previously assigned Serving GW/PDN GW(s) (this part is not shown in the graph, for simplicity).
11. The MME sends the NAS signaling message "TAU Accept" to the UE; it can assign a new temporary id.
12. If a new temporary identity was assigned to the UE in the previous message from the MME, the UE confirms that back to the MME by a "TAU Complete" message.

5.8.2 Intra-E-UTRAN Handover via X2

This is the most efficient and performant handover variant, and expectedto be the most frequent case. A prerequisite is X2 connectivity between old (source) and new (target) eNodeB. The message flow below (see Figure 5-20) concentrates on the signaling steps involving the EPC entities for handover completion, and those within the E-UTRAN (handover preparation and execution) are explained only in brief. For simplicity it does not contain a Serving GW change; the difference would be (1) usage of Create Session request/response with new Serving GW instead of Modify Session request / response with the old Serving GW, and additionally (2) deletion of session between MME and old Serving GW.

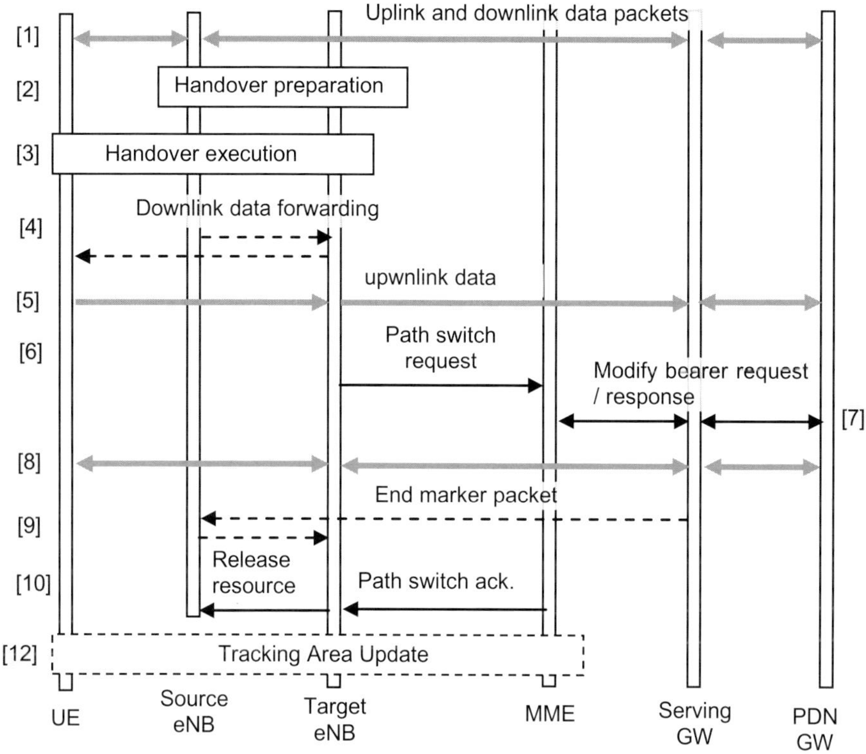

Figure 5-20: Intra-E-UTRAN handover via X2

We see the following steps being executed (steps 2 and 3 happen fully in the radio access network and we do not explain the details here):

1. Uplink and downlink data packets are flowing between UE and PDN GW via source eNodeB and Serving GW.
2. Handover preparation between source and target eNodeB is performed via X2.
3. Handover execution between source and target eNodeB is performed via X2.
4. As a result of handover, forwarding of downlink data packets from the source eNodeB to the target eNodeB may occur (if lossless handover is required).
5. Uplink data travels already via the target eNodeB.
6. The target eNodeB requests the switch of the traffic path from the MME
7. MME performs update signaling with Serving GW by exchanging an Modify Bearer request/response message pair; optionally the location information may be included. MME could also have released bearers; this in turn may lead to an update to the PDN GW about the bearer properties. If the location in-

formation was included by MME, it is also propagated upwards to PDN GW where it can be used for charging purposes.

8. The traffic now flows via the target eNodeB in both uplink and downlink direction.
9. Immediately after switching the traffic path, the Serving GW sends one or more end-marker packets in the user plane to the source eNodeB, from where they are forwarded to the target eNodeB. This helps in the reordering process of data packets.
10. The Serving GW achnowledges the path switch back to the MME; MME instructs the source eNodeB to release the resources.
11. If necessary, i.e. if the UE entered a new Tracking Area, the UE performs a Tracking Area Update (NAS signaling message to MME relayed via the target eNodeB, see sub-section 5.8.1).

5.8.3 Intra-E-UTRAN Handover via S1

The somewhat lengthy handover flow is split here into preparation phase (Figure 5-21) and execution (Figure 5-22) and applies if there is no X2 connectivity between source and target eNodeB. In this example signaling flow also MME and Serving GW relocations are included. The graph is simplified, e.g. a status notification from source to target eNodeB via MMEs is left out between steps 10 and 11; it applies only if PDCP status preservation is necessary. Also, the ISR feature is not elaborated here. Only one PDN GW is shown, but the extension for the multiple PDN case is quite obvious.

During the handover preparation phase all signaling is performed to find the new nodes and inform them about the pending handover, exchange context confirmation and to prepare bearers and, optionally, the forwarding data path. These are the steps up to the point just before the UE changes the radio access to the cell of the target eNodeB.

The following steps are performed in this procedure:

1. Data packets are flowing downlink and uplink over the old traffic path (i.e. via source eNodeB and source Serving GW).
2. The source eNodeB decides, e.g. due to lacking connectivity with the target eNodeB or as a result of a failure of handover via X1, that a handover via S1 is required.
3. The source eNodeB sends a "handover required" message to the source MME, indicating whether direct forwarding is to be supported, target eNodeB and target TAI. Also a transparent container with data to be transferred from source eNodeB to target eNodeB is included.

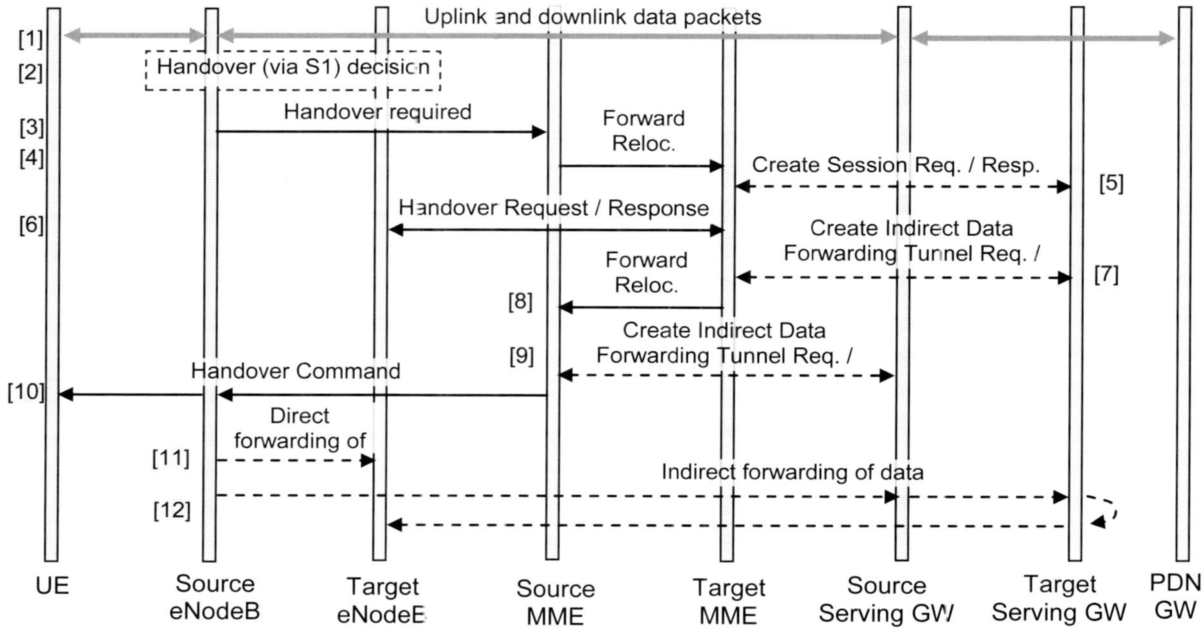

Figure 5-21: intra-E-UTRAN handover via S1 (preparation part)

4. The source MME determines the target MME and sends a Forward Relocation request to it, including UE's current MME context and the relevant information received from the source eNodeB.
5. The target MME checks whether the Serving GW can still be used; in this example message flow it is not the case, and a new, target Serving GW is determined. Then the target MME exchanges a Create Session message pair with the target Serving GW, with information like bearer contexts, PDN GW addresses, tunnel endpoint ids/GRE keys at the assigned PDN GWs, for uplink traffic. The Serving GW also assigns its own tunnel endpoint identities for uplink traffic.
6. The target MME exchanges a Handover Request / Acknowledge message pair with the target eNodeB, including the previously received transparent container and information on bearers. This creates the UE context and security context in the target eNodeB. Also tunnel related information for downwlink and forwarded traffic is exchanged, and in the response from the target eNodeB another transparent container (from target to source eNodeB) plus eventually rejected bearers are included.
7. If indirect forwarding applies, the target MME exchanges a message pair with the target Serving GW; again, tunnel related information (this time for forwarding) is transferred between the two nodes.
8. The target MME sends all relevant information (e.g. on the forwarding tunnel, bearers, Serving GW address and the transparent container from target to source eNodeB) to the source eNodeB.
9. If indirect forwarding applies, the Source MME exchanges a signaling message pair with the source Serving GW for creation of the forwarding tunnel between source and target Serving GW; the information which bearers are subjet to forwarding is included.
10. The source MME sends a handover command to the source eNodeB, including target to source container and bearer information (which bearers to forward and which ones to release), from where it is relayed to the UE.
11. If direct forwarding applies, the source eNodeB starts forwarding downlink data packets to the target eNodeB via X2.
12. If indirect forwarding applies, the source eNodeB starts now forwarding of downlink data packets to the target eNodeB via source and target Serving GWs.

In case the handover fails at step 6 (i.e. the target eNodeB cannot allocate resources for any bearers), the reservations made so far in step 5 need to be released again, and a handover preparation failure indication is finally sent to the source eNodeB (this part is not shown here).

The procedure continues with the handover execution part (see Figure 5-22).

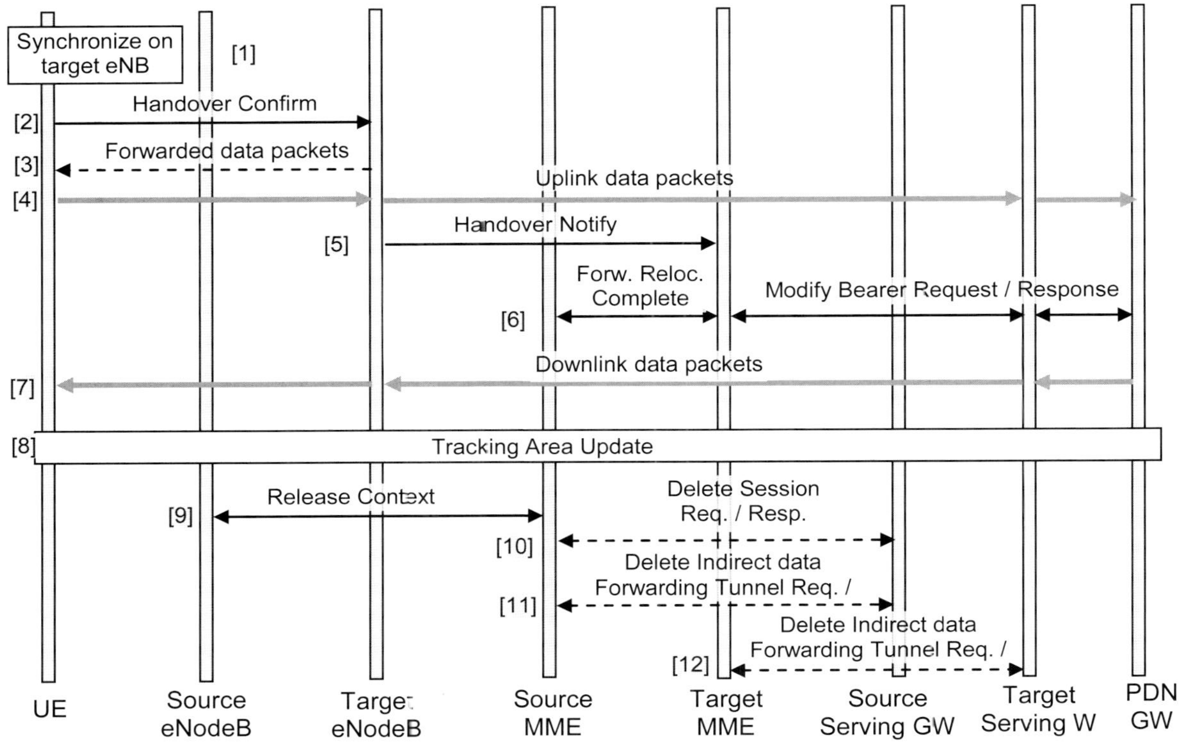

Figure 5-22: intra-E-UTRAN handover via S1 (execution part)

1. The UE synchronizes with the radio channels of the target eNodeB.
2. The UE sends a message to confirm its handover to the target eNodeB.
3. Data packets forwarded from the source eNodeB can now be delivered to the UE.
4. UE's uplink data packets will now travel on the new traffic path (via target eNodeB and target Serving GW).
5. The target eNodeB botifies the target MME of the successful handover. It contains the identities of the new tracking area and cell.
6. The target MME exchanges a pair of signaling messages with the source MME to inform about the success of the handover. In the source MME a timer is started to supervise the teardown of resources in source eNodeB and source Serving GW; in target MME a timer is started to supervise the resources for forwarding (if applicable). Also, the target MME updates the traffic path by exchanging a Modify Bearer request/response pair with the target Serving GW, which does the same with the PDN GW. Main information here is the bearer properties and tunnel properties. In case of multiple PDN connectivity this has to be repeated for every PDN connection.
7. Now also downlink data flows solely over the new path (via target Serving GW and target eNodeB).
8. If necessary, according to the defined triggers, a tracking area update in a simplified manner is performed (leaving out some unnecessary steps, like fetching of UE context from the source MME).
9. After the timer started in step 6 has expired, the source MME exchanges with source eNodeB a message pair to let it release UE's context.
10. After the timer started in step 6 has expired and source MME has also received the Forward Relocation Complete message (of step 6), it exchanges a pair of signaling message swith the source Serving GW to let tear down Serving GW's resources (EPS bearers established till now with the PDN Gateway, but due to the indication that the Serving GW has changed, these are not completely deleted at the PDN Gateway!). (This step applies in this example message, but is shown with dashed line, because it is done generally only if the Serving GW changed).
11. The source MME also instructs the source Serving GW to tear down the resources for indirect forwarding (if such forwarding was used).
12. The target MME instructs the target Serving GW to tear down the resources for indirect forwarding (if used).

5.9 Inter-System Mobility with Legacy 3GPP

Four handover cases exist between E-UTRAN and UTRAN/GERAN, two per direction and for both Iu mode and A/Gb mode. We limit our presentation to the E-UTRAN to UTRAN Iu mode case and divide the handover message flow for convenience again into preparation phase (Figure 5-23) and execution phase (Figure 5-24). GTP is assumed on the S5/S8 interface. Either direct data forwarding (between eNodeB to RNC) or indirect data forwarding (from eNodeB to the new Serving GW via the old Serving GW) can be used; note that in the latter case data packets pass the same (old) Serving GW twice, once as a normal downlink packets, and a second time as forwarded data packets.

The UE is in connected mode at the start of the handover. In the preparation phase the resources in the target system (between new SGSN and new Serving GW) are prepared, and if the MME has this feature configured, also for indirect downlink data forwarding during the handover (between source Serving GW and target SGSN). Only one PDN GW is shown, in case of multiple PDN connections several tunnels (between Serving GW and respective number of PDN GWs) need to be handled.

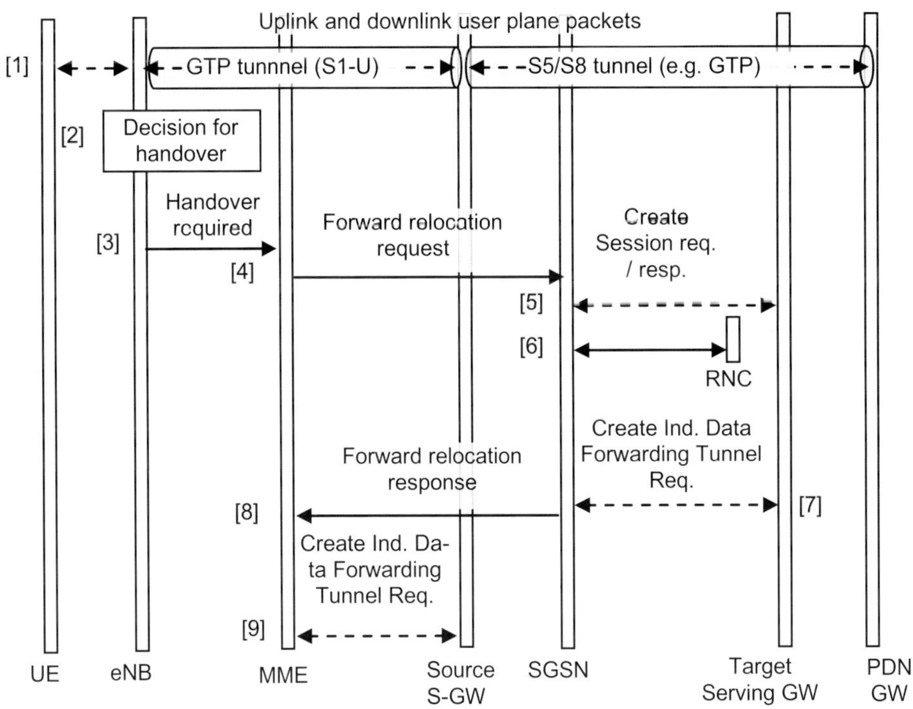

Figure 5-23: E-UTRAN to UTRAN Iu mode handover (preparation phase)

1. The user plane is installed between eNodeB and Serving GW (GTP tunnel) and between source Serving GW and PDN GW (e.g. GTP), and IP packets may flow in uplink and downlink.
2. The eNodeB decides (based on measurement reports from the UE) that handover to the GERAN/UTRAN is required.
3. The eNodeB sends a request for handover to the MME. Information on the target RNC is included.
4. The MME selects a (target) SGSN and sends a forward relocation request message to it.
5. The (target) SGSN exchanges signaling (message pair "Create Session request / response") with the target Serving GW (only needed if the Serving GW is changed).
6. The (target) SGSN requests the establishment of resources in the target radio access network (RABs in UTRAN).
7. If indirect data forwarding is configure, the (target) SGSN sends a request for establishment of an indirect data forwarding tunnel to the target Serving GW.
8. The (target) SGSN acknowledges the forwarding relocation request of step 4 back to the MME.
9. If indirect data forwarding is configured, the MME instructs the source Serving GW to establish an indirect data forwarding tunnel.

The procedure continues then with the actual execution of the handover, according to Figure 5-24, with these steps:

1. The user plane is established in the source system (GTP tunnel between source eNodeB and source Serving GW, and GTP or PMIP tunnel, depending on the EPC protocol variant, between Source Serving GW and PDN GW). Data packets can flow from/to the UE in uplink and downlink direction.
2. The MME issues the handover command to the (source) eNodeB, which sends the "Handover from E-UTRAN" message to the UE.
3. The UE performs the access procedure with the UTRAN radio access network (for simplicity NB and RNC are not shown separate here).
4. To indicate the completion of access to UTRAN, the UE sends a corresponding message to the UTRAN (via NodeB to RNC).
5. At this point uplink data packets can be sent from the UE through the target (UTRAN) access; however, the user plane tunnels in the core network are not yet established, so they need to be buffered at the RNC.
6. Downlink data packets are still arriving through the source access, and are forwarded, if configured, from the source eNodeB. Two variants for forwarding are possible: a) directly to the target RNC or b) via source and target Serving GW and via target SGSN to the target RNC. (The target SGSN is skipped if Direct tunnel scheme is applied in the target system; in this case the forwarding tunnel utilizing S12 is done directly from the target Serving GW to the target RNC.)

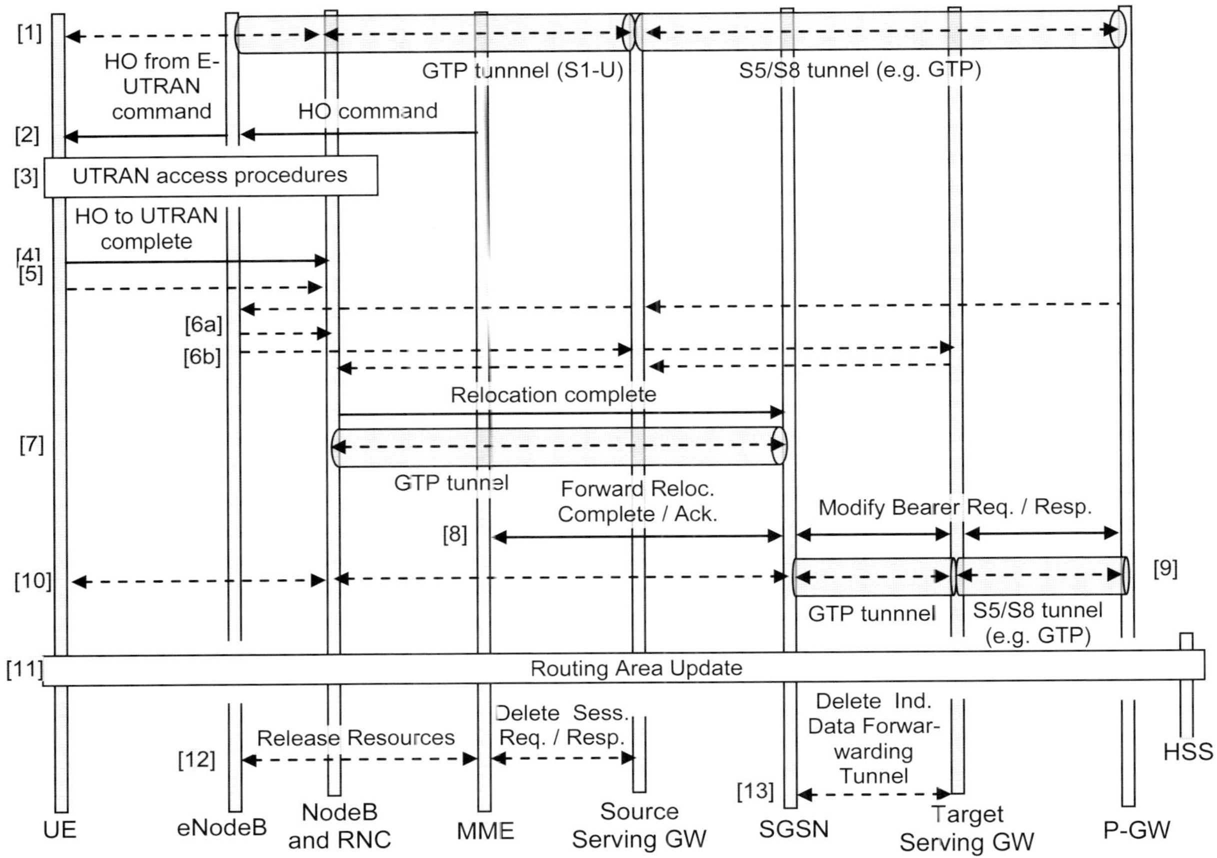

Figure 5-24: E-UTRAN to UTRAN Iu mode handover (execution phase)

7. The target RNC informs the (target) SGSN that the handover is completed in the UTRAN by sending a “Relocation complete” message (the procedure resembles the RNC relocation within UTRAN). As a result, the GTP tunnel for the user plane is established.
8. The (target) SGSN exchanges a message pair with MME to indicate that the UE has arrived in the target system; the MME starts a timer for supervision of resources in the source system. When receiving the acknowledgement from MME, the (target) SGSN starts also a timer for supervision of release of resources for indirect data forwarding (if it was applied).
9. The (target) SGSN exchanges signaling messages with the target Serving GW for establishment of the resources (GTP tunnel). The target Serving GW does the same, correspondingly with the PDN GW for establishment of the S5/S8 resources; here the tunnel could be based on GTP or PMIP, depending on the EPC protocol variant.
10. Data packets can now flow in both directions through the target system.
11. The UE performs a Routing Area Update procedure in the target system, involving RNC, (target) SGSN, MME and HSS.
12. After expiry of the timer set in step 8, the MME sends a message to the source eNodeB to release the resources. Also, a message is sent to the source SGSN to delete the session for the UE and related resources with the source Serving GW (this step is only needed if the S-GW has changed during the handover).
13. When the timer previously set in step 8 expires, the target SGSN releases the indirect data forwarding tunnel resources with the target Serving GW.

5.10 Inter-System mobility with non-3GPP access

Many different handover cases exist, depending on the type of the non-3GPP access and the mobility protocol used. The direction of the handover multiplies the number by 2 (or rather almost, minus the cases where source and target access are of the same type).

5.10.1 Handover from E-UTRAN to Trusted non-3GPP Access

This is a non-optimized handover flow, and two variants exist, depending whether GTP or PMIP is used in the EPC. The difference lies in the handling of policies with the user plane and corresponding signaling: this is combined within GTP, and separated into PMIPv6 handling and PCC message flows). E-UTRAN and MME are not shown, for simplicity and because they are not involved in the procedure.

The overall message flow is shown in Figure 5-25.

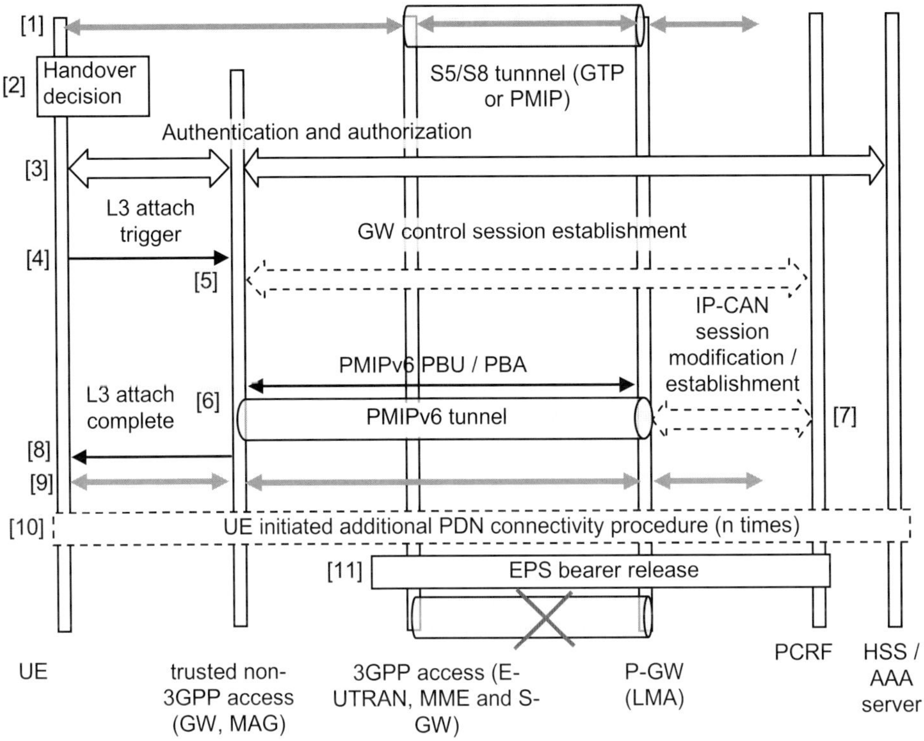

Figure 5-25: message flow for handover from E-UTRAN to trusted non-3GPP access

Assuming again a co-location of functions in the trusted non-3GPP access network (e.g. as represented by an access gateway), the steps are in detail:

1. The UE is in 3GPP access (E-UTRAN or GERAN/UTRAN) and the S5/S8 tunnel is established (either with GTP or PMIP). Data packets may flow in both directions.
2. The UE has discovered the non-3GPP access (e.g. by scanning according to network discovery information received from ANDSF previously) and decides to perform handover (e.g. based on inter system mobility policies received from ANDSF, or stored locally).
3. The UE performs access authentication and authorization (as already explained in sub-section 5.4.3) for the trusted non-3GPp access: as a result, the PDN-GW identity (or identities, if the UE is connected to multiple PDNs) is conveyed to the trusted non-3GPP access, so that the MAG knows where to send subsequent mobility signaling messages. The AAA server also returns to the non-3GPP access the UE identity in form of a NAI, which is to be used in subsequent mobility signaling (it is derived from the IMSI).

4. The trigger for attachment at layer 3 is sent from the UE to the trusted non-3GPP access; this is done in an access specific way.
5. If use of dynamic policies is configured between the 3GPP network and the trusted non-3GPP access network (i.e. the interface Gxa is supported), the messages for establishing the GW control session are exchanged between the non-3GPP access network and the PCRF.
6. The PMIPv6 mobility signaling (Proxy Binding Update / Proxy Binding Acknowledge messages) is exchanged between MAG function in the trusted non-3GPP access network and the LMA in the PDN-GW (for details see sub-section 4.15.2). The MAG uses the UE identity as received from the AAA server in step 3. In the response from the PDN-GW, assuming that IP address preservation is granted by corresponding IP mobility mode indication in AAA signaling (step 3), the same IP address information (IPv4 address or IPv6 prefix) as already used in the 3GPP access is returned to the UE; if the IP address preservation is not granted, new IP address information is sent instead. An PMIPv6 tunnel is installed between the LMA and the MAG.
7. If dynamic PCC is employed, then if the IP address was preserved, the PDN-GW updates the IP-CAN session with the PCRF; if the IP address was not preserved, a new IP-CAN session is established.
8. The completion of attachment on layer 3 is signaled from the trusted non-3GPP access network to the UE (in an access specific way).
9. Now data packets can flow end-to-end to/from the UE via the trusted non-3GPP access.
10. If the UE was connected to multiple PDNs before the handover, it can execute the procedure for additional PDN connectivity with other PDN-GWs (as many times as required).
11. The PDN-GW releases the EPS bearers, by executing the PDN disconnection procedure or the bearer deactivation procedures.

5.10.2 Handover from E-UTRAN to Untrusted non-3GPP Access

The message flow, visible in Figure 5-26, is similar to the previous one, with these characteristic differences (see also the procedure “initial attach in untrusted non-3GPP access”, sub-section 5.4.4):

1. No GW control session is established between ePDG and PCRF (the provisioning of dynamic policies to ePDG is not supported in 3GPP Rel. 8).
2. Instead of access specific signaling and user plane establishment, IKEv2 signaling and IPSec as a tunnel for the user plane are used. The L3 attach trigger is thus conveyed by IKEv2 signaling.

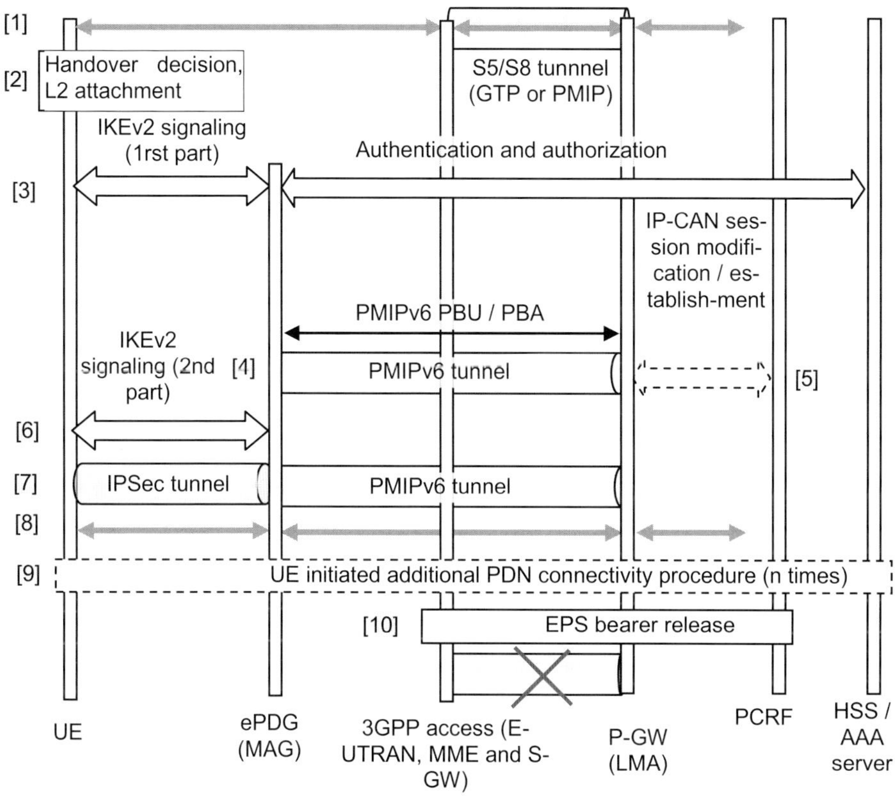

Figure 5-26: message flow for handover from E-UTRAN to untrusted non-3GPP access

Steps 1 and 2 are the same as in the handover from E-UTRAN to Trusted non-3GPP access (Figure 5-25); additionally the UE has to perform local IP address configuration within the untrusted non-3GPP access. The subsequent steps 3 and 8 in the procedure are similar to those in sub-section 5.4.4 and need not be repeated here in full. However, in step 4 it depends on the IP mobility mode chosen, whether the PDN-GW will convey new IP address information or confirm the current one.

Steps 9 and 10 are again the same as steps 10 and 11 in the handover from E-UTRAN to Trusted non-3GPP access (Figure 5-25).

5.10.3 Handover between two Untrusted non-3GPP Accesses

The procedure is based on MOBIKE [3] (see sub-section 6.4 for protocol details). The condition is that the same ePDG can be maintained also when connecting to the EPC via the new non-3GPP access network. (Note: this may not always be the case, due to restrictions from roaming agreements or due to traffic routing strategies.) Also, the ePDG needs to support the MOBIKE capability and aware of its use, which is indicated in the establishment of the IPSec tunnel e.g. in the initial attach.

The handover optimizes the handover signaling, and is realized by the following steps (see Figure 5-27):

1. IPSec and PMIP tunnels exist between the UE and ePDG, and ePDG and PDN GW, respectively. The UE also has been allocated a remote IP address from ePDG (which was used for establishing the PMIPv6 tunnel towards PDN GW).
2. The UE dis-associates at layer 2 from the source untrusted non-3GPP access network.
3. The UE associates at layer 2 with the target untrusted non-3GPP access network.
4. The UE re-configures its IP layer, i.e. receives and uses an IP address from the target untrusted non-3GPP access network.
5. The UE exchanges a roundtrip signaling message (MOBIKE extension of IKEv2) to inform the ePDG about the new IP address.
6. Optionally, the ePDG may verify that new IP address of the UE by a MOBIKE defined roundtrip test message.
7. The ePDG updates the tunnel configuration, and also its internal mapping from the PMIPv6 tunnel.
8. The IPSec tunnel is updated, the PMIPv6 tunnel remains unchanged (thanks to the fact that the remote IP address has not changed).

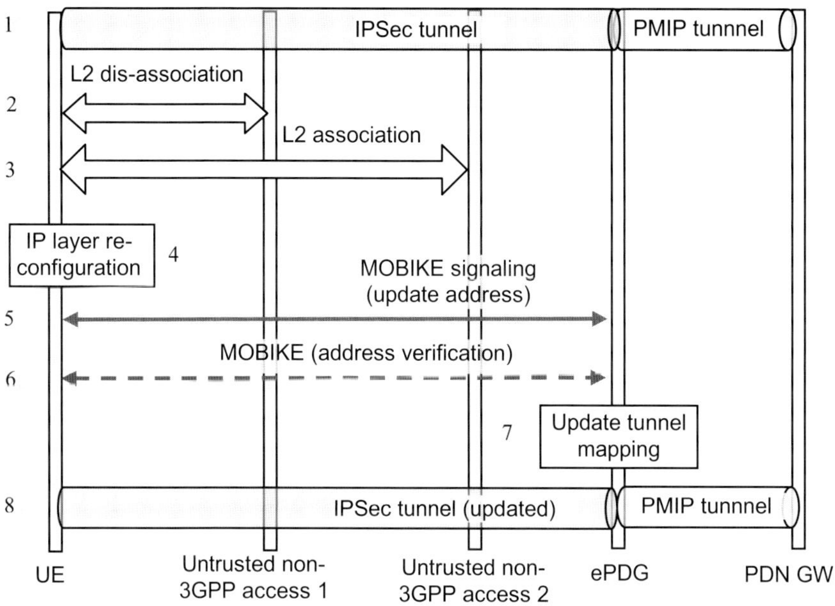

Figure 5-27: message flow for handover between two untrusted non-3GPP accesses (same ePDG)

The savings in signaling and handover latency are appreciable, compared to the non-optimized case, where the IPSec tunnel would have to be torn down and re-created anew through extensive IKEv2 signaling.

5.10.4 Optimized Handover between CDMA 2000 © HRPD and E-UTRAN

The procedure is applicable for a dual (E-UTRAN/HRPD capable) mode UE, if the involved networks are configured appropriately, and is divided into three stages:

1. Pre-registration (Figure 5-28);
2. Handover preparation (Figure 5-29);
3. Handover execution (Figure 5-30).

Pre-registration can take place at some point in time before the handover. The UE stays in E-UTRAN and the necessary HRPD related signaling is tunneled through S101. This tunneling interface must be configured between those networks that wish to support the optimized handover.

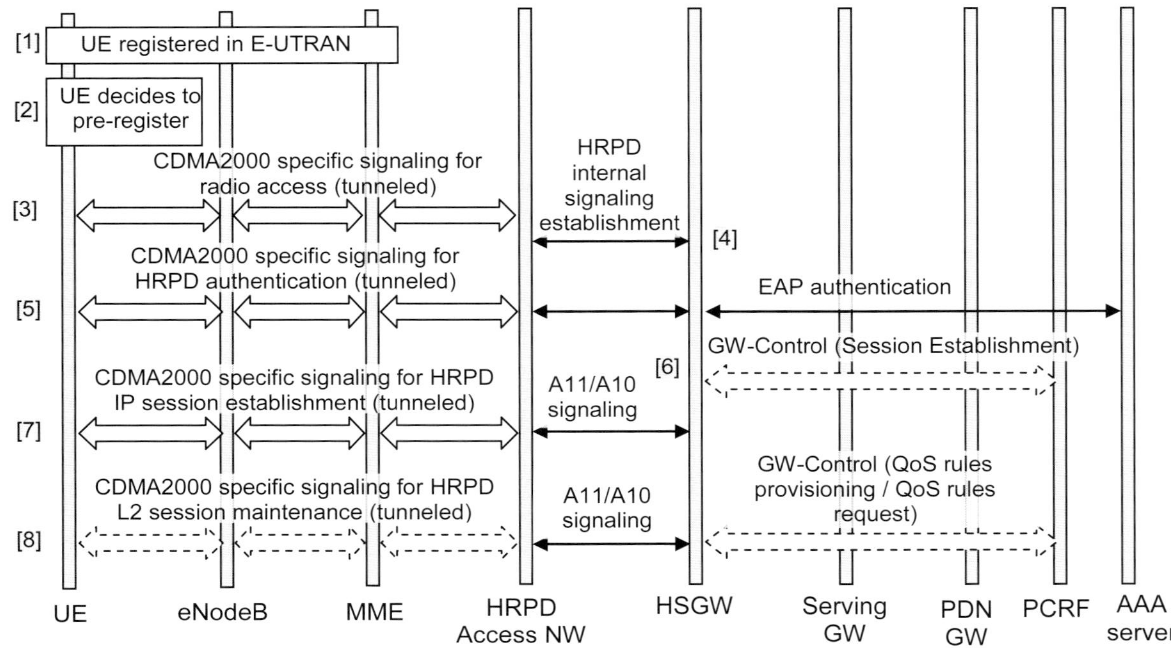

Figure 5-28: message flow for optimized handover from E-UTRAN to CDMA2000 © HRPD (pre-registration)

The message flow for pre-registration extends over 9 nodes, but two of them (eNodeB and MME) forward these messages only, without interpretation; it consists of these steps in detail:

1. The UE is registered in E-UTRAN and is either in idle or active mode.
2. Based on information received from E-UTRAN in step 1 (broadcast in idle mode or explicit indication from eNodeB in active mode), the UE decides that it may be advantageous, for the sake of subsequent handovers, to pre-register with the HRPD system.
3. The UE exchanges signaling with the HRPD access network for establishing a session on radio level/L2; the details or meaning of these are out of scope here, but it is important to note that eNodeB adds the HRPD related information "Sector ID" to the HRPD specific message, which is carried in an unstructured container. Sector ID is statically configured in the eNodeB and used by the MME to address the S101 termination node in the HRPD access network. MME assigns a S101 session ID and tunnels the HRPD signaling to this node. All three message transfers are acknowledged backwards, with reference to the session ID.
4. HRPD internal signaling is performed between the access network and the HSGN.
5. EAP authentication is performed for the HRPD access network (as a trusted non-3GPP access) between UE and AAA server. This is an HRPD specific L2 transport between the node in the HRPD access network and the UE, again tunneled via the E-UTRAN and MME to the HRPD access network, where it is forwarded to the HSGW, which acts as the authenticator. The backend part of AAA signaling is likely via an AAA proxy (not shown in the figure) and based on STa interface using DIAMETER (see sub-section 6.5). Assigned APN and identity of the PDN GW are sent back, comparable to initial attachment in (generic) trusted non-3GPP access (sub-section 5.4.3).
6. Optionally, if dynamic PCC is employed, the GW-control session is established between HSGW and PCRF. For bearer binding, the PCRF needs to provide the QoS rules for all bearers to be handed over, so that the HRPD system can establish them.
7. HRPD internal signaling, and the forwarding towards the UE (tunneled via the EPC and E-UTRAN) for IP session creation is performed. Previously received QoS rules will be propagated in downlink direction (i.e. the equivalent to EPS-bearer and RAB creation happens in HRPD nodes, including UE).
8. If the HRPD session context need to be changed from either side (HRPD access/HSGW or PCRF in the EPC), corresponding signaling is executed. QoS rules are pulled by HSGW or pushed down by PCRF.

The UE may fall into (E-UTRA/EPC) idle state after step 7. For step 8, it needs to moveinto active state. In HRPD, the state is dormant.

The handover preparation phase is visualized in the Figure 5-29; prior to that, the above pre-registration must have successfully been performed. A GTP forwarding tunnel from eNodeB to the Serving GW, and the S103 tunnel between Serving GW and HSGW are used if forwarding of downlink data should occur during the handover (per S1 bearer, as long as the downlink data path in the target system/HRPD is not yet ready). PDN GW, PCRF and HSS are not involved in this procedure.

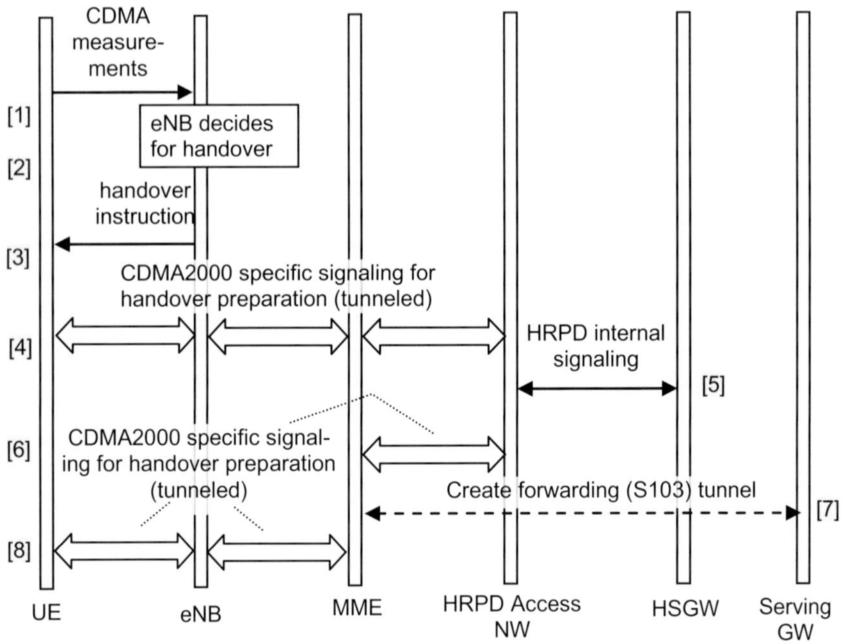

Figure 5-29: message flow for optimized handover from E-UTRAN to CDMA2000 © HRPD (handover preparation phase)

The handover execution phase consists of these steps (as a prerequisite the UE is connected via E-UTRAN to EPC, and not e.g. via 2G/3G access):

1. The UE delivers CDMA 2000 © radio measurements, according to the configuration imposed by the eNodeB.
2. The UE decides for a handover; the decision would be based on signal strengths of the current E-UTRAN cell and neighbouring CDMA 2000 © HRPD cells.
3. The eNodeB sends an instruction to perform handover to CDMA 2000 © HRPD (by a corresponding RRC message).

4. The UE requests access to radio channel from HRPD, and indicates that this due to pending handover from E-UTRAN; this message is tunneled through the eNodeB (which adds Sector ID) and EPC (MME) to the HRPD access network. Important information added by MME are one or more of the items PDN GW Identity, GRE key for uplink traffic and APN; the GRE key(s) is/are used for the S103 forwarding tunnel between MME selects the node in HRPD access according to the Sector ID and memorizes a session id.
5. The HRPD access network allocates the resources and sends APN(s), PDN GW identity(ies) and uplink GRE key(s) to the HSGW by HRPD internal signaling. The response message from HSGW includes downlink GRE key(s) for forwarding traffic on S103.
6. The HRPD specific signaling is tunneled to the MME; also additional information needed for MME is included, like status of the handover procedure and, if forwarding is to be applied, S103 tunnel properties (HSGW address, GRE key(s)).
7. If data forwarding applies, the MME determines for which EPS bearers, and instructs the Serving GW with the necessary information. The Serving GW confirms the resources for forwarding via S103 back to MME, including uplink tunnel endpoint IDs (one per forwarded S1 bearer).
8. The MME conveys the information about Serving GW address and uplink tunnel IDs to the eNodeB for use in the subsequent forwarding process, and includes a cause (on success or failure).The eNodeB finally instructs the UE to perform the handover, encapsulating the original response from HRPD system and related access information. At this point the forwarding of data is started by the eNodeB.

The second part, in Figure 5-30, comprises the actual execution of the handover:

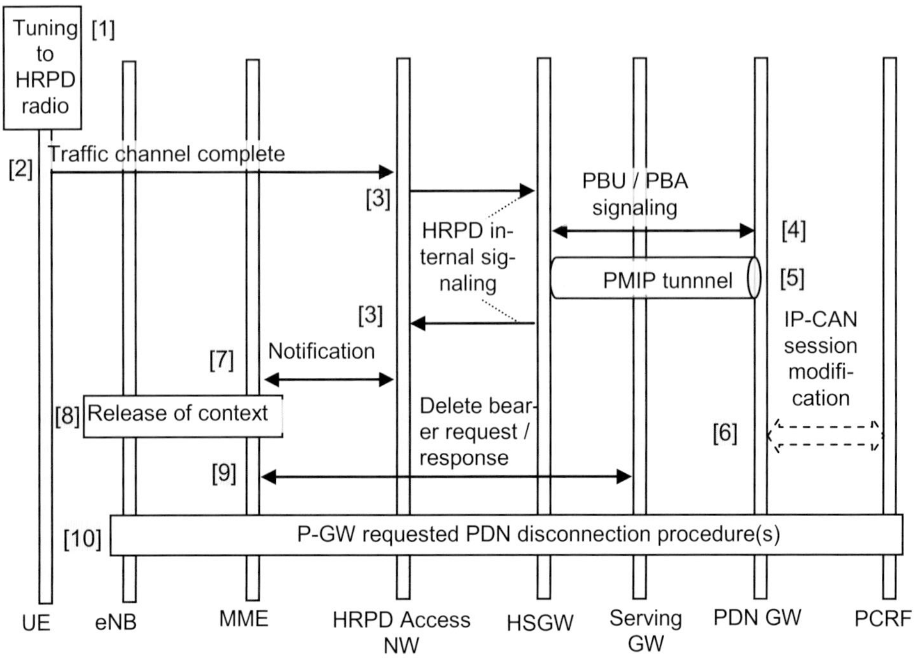

Figure 5-30: message flow for optimized handover from E-UTRAN to CDMA2000 HRPD (handover execution phase)

1. The UE tunes to the HRPD radio signal and acquires the traffic channel indicated within the preparation signaling.
2. The UE sends a message to the HRPD access network to indicate that the traffic channel selection has been performed successfully.
3. HRPD internal signaling is performed between HRPD access network and HSGW (request and confirmation). This installs the user plane.
4. The HSGW executes PMIP signaling with PDN GW (Proxy Binding Update and Proxy Binding Acknowledge messages), to install the mobility binding in PDN GW. No new IP address information is requested, so that the PDN GW confirms the existing IPv4 address and/or IPv6 prefix.
5. The traffic is switched and the PMIP tunnel is now in place.
6. If dynamic PCC is used, the P-GW sends a request for policy rules to the PCRF, indicating the change in IP-CAN (can be done in parallel to the PBA message of step 4). Otherwise, static policies configured in P-GW are used.
7. The HRPD access network sends a notification to the MME, indicating that the HO is completed. The MME sets a timer for resource release in EPC.

8. The context in E-UTRAN is released.
9. If the timer(s) in MME for resource supervision expires, MME instructs the Serving GW to release S1 bearers (if not yet happened) and resources for forwarding (S103 tunnel).
10. The PDN GW triggers the release of the resources in E-UTRAN and S1 bearers. This can happen at any time after step 5.

For multiple PDN connections, steps 4, 5 and 10 need to be repeated per PDN connection.

5.11 Session handling procedures

These procedures deal with the establishment, modification and deletion of EPS bearers between UE and PDN GW (which includes Radio Bearer, Radio Access Bearer/S1 bearer and S5/S8 bearer) and the interaction with PCC (if used). Note that a default bearer is activated with the attachment to every PDN.

Dedicated bearer activation is always triggered by the network (but it can be due to UE requested bearer resource modification). Modifications of bearers can be requested by the UE or by the network e.g. due to

- application level signaling e.g. from P-CSF to the PCRF, or a policy change through PCRF or in PDN GW;
- subscription change in HSS;
- radio resource problems by the E-UTRAN (signaled via MME);
- non-optimized handover by PDN GW, when all EPS-internal resources are released, except the IP address).

In the following we describe a few illustrative cases.

5.11.1 Dedicated Bearer Activation

The message flow in Figure 5-31 indicates the signaling for creation of a new bearer, for GTP as S5/S8 tunneling protocol; precondition is that a connection with the target PDN (and thus a default bearer) already exists.

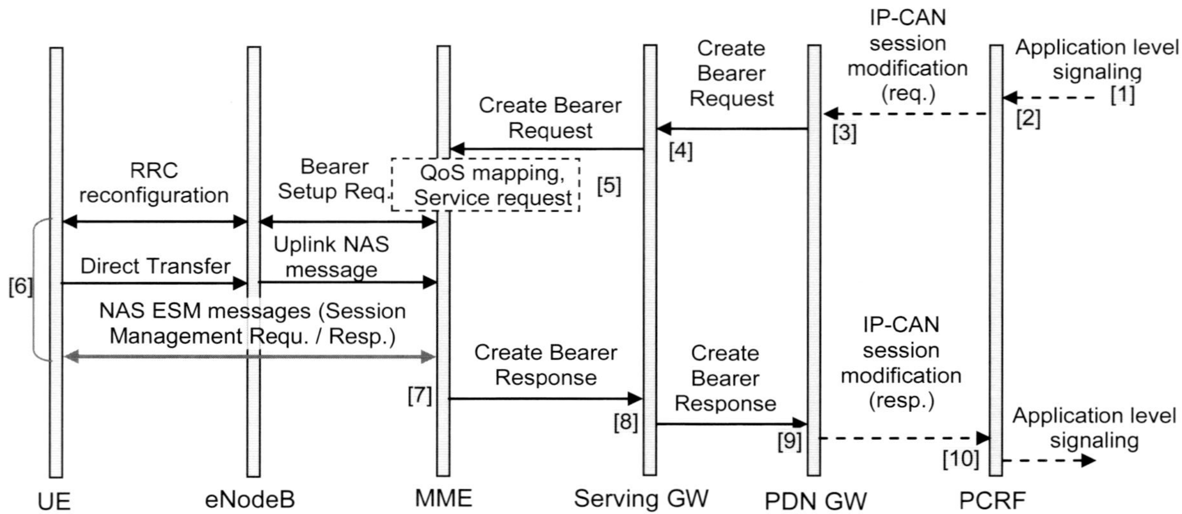

Figure 5-31: dedicated bearer activation procedure

1. Application level signaling, indicating a need for dedicated resources, arrives at the PCRF (e.g. through the Rx interface from the P-CSCF, triggered by SIP/SDP signaling between UE and IMS). Alternatively, a PCRF-internal trigger event related to QoS policies may occur.
2. If dynamic policies are deployed: PCRF sends a request for IP-CAN session modification (via DIAMETER messages Re-Auth-Request of the Credit Control Application) to the PDN GW, including information on the QoS.
3. Alternatively to step 2, the PDN GW may employ a local QoS policy for bearers, which requires the creation of a new dedicated bearer. The PDN GW sends a request for creating a dedicated bearer to the Serving GW (via GTP-C), including information on the UE identity (IMSI), linked bearer id (which identifies the associated default bearer), EPS bearer QoS and Traffic Flow Template.
4. The Serving GW forwards the request for creating a dedicated bearer to the MME.
5. If the UE is also capable for GERAN/UTRAN, the MME derives the legacy QoS parameters. If the UE is in idle mode, the MME executes a network triggered service request procedure.
6. The MME sends the S1-AP message a "Bearer Setup request" with all necessary parameters to the eNodeB, but excluding ARP parameter (as it is for its internal use). The eNodeB maps the EPS bearer QoS to the radio bearer QoS and sends a RRC Reconfiguration message to the UE. The NAS related session information is included in these two messages and is delivered to the UE, which stores the uplink TFT for mapping the traffic flows onto radio bearers. The UE may indicate the EPS bearer QoS parameters also to applications. The messages are acknowledged back to the senders with an indication whether the EPS QoS could be allocated (again ESM session information from UE to MME is included RRC message "Direct Transfer" and S1-AP message "Uplink NAS Transport").
7. The MME acknowledges the activation of the bearer by sending a Create Bearer Response message to the Serving GW.
8. The Serving GW in turn sends the Create Bearer Response message to the PDN GW.
9. If the bearer activation was triggered by PCRF, the PDN GW informs PCRF thereof and indicates, whether the request for resources could be fulfilled.
10. If the resource request was due to application level signaling, e.g. from P-CSCF, the completion and result of the resource reservation in EPC and E-UTRAN is signaled back at the end of the procedure.

5.11.2 Bearer Modification (incl. QoS update)

This procedure (Figure 5-32) is used if one of the QoS parameters QCI, GBR, MBR or ARP is modified e.g. by PCRF due to dynamic policies; it is partially used if the subscription profile is changed in HSS (in this case corresponding update signaling would be from HSS to MME, which informs Serving GW and PDN-GW). Note that a modification from a QCI type non-GBR (no guaranteed bit rate) to GBR (guaranteed bit rate) is not supported. Figure 5-32 shows the bearer modification procedure with bearer QoS update. If the bearer QoS is not to be updated, the procedure is similar, but without S1/RRC related reconfiguration messages; instead only plain NAS related signaling messages are exchanged between MME and UE (via the eNodeB).

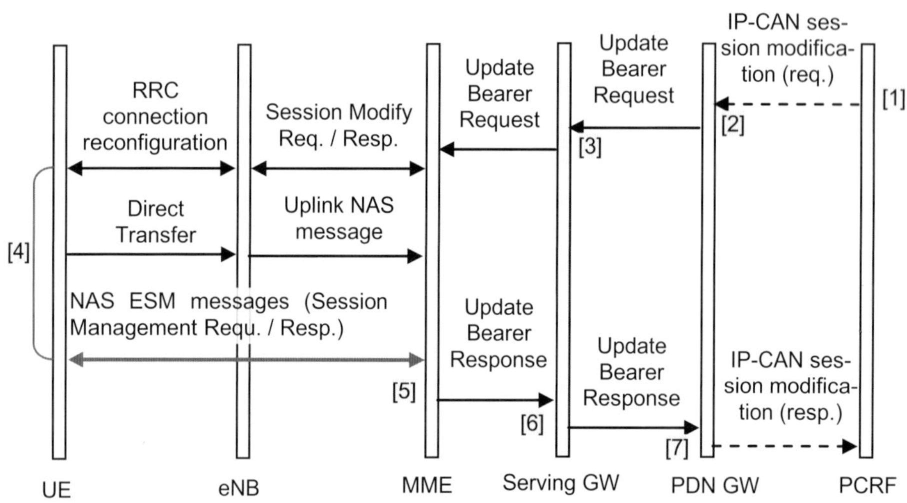

Figure 5-32: bearer modification procedure

From the appearance of the flow, this procedure has much in common with the dedicated bearer activation procedure (see before); message names are partially different, and further deviations are:

1. The PDN GW determines from the QoS policy received from PCRF (if dynamic PCC is employed) or from its own static QoS rules that the authorized QoS of a service data flow needs to be changed, or that a service data flow is to be aggregated to or removed from an active bearer. The PDN GW generates the traffic flow template, updates the EPS Bearer QoS and sends the Update Bearer Request to the Serving GW.
2. If the APN AMBR has changed, the MME updates the UE AMBR (if still within the limits of the subscriber data).
3. The UE may notify the change in QoS parameters to applications.

5.12 Communication between UE and ANDSF

The UE may receive intersystem mobility policy data and data for access network discovery in push or pull mode from the ANDSAF server. Initially the UE has to discover the ANDSF and establish a secure connection. An example message flow for push and pull of ANDSF policies is given in Figure 5-33. Care has to be taken that this signaling, especially the pull for access network discovery data by the UE, does not happen too often; also, it was explicitly meant not to couple it with the actual handover signaling. Actually this requires a pro-active, forward-looking behaviour of ANDFS and UE, where data is pushed or pulled well before it is urgently needed for a concrete handover.

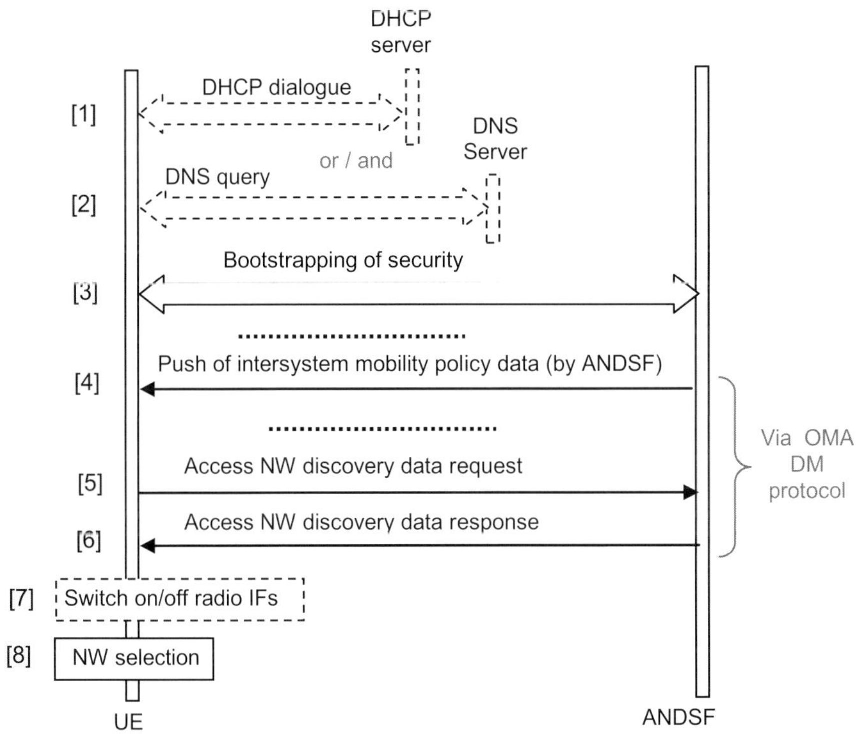

Figure 5-33: example message flow for communication between UE and ANDSF

The necessary steps are:

1. The UE issues a DHCP query, indicating ANDSF IP address or Domain name List as required information; IETF has defined the options for both DHCPv4 and DHCPv6. The response is thus either an IP address of the requested IP version or a FQDN.

2. If the response from DHCP was a FQDN, or as an alternative to step 1 (if the UE has a FQDN for ANDSF preconfigured), the UE resolves it by a DNS query.
3. As a result of steps 1 and/or 2, or by preconfigured data, the UE has now available an IP address for the ANDSF server. It bootstraps a security association with the ANDSF server, based on the generic authentication architecture framework of 3GPP (see 3GPP TS 33.222 [5]), which supports https; now the transfer of data by OMA DM means via S14 interface is protected.
4. At some later point and based on a network internal trigger, the ANDSF pushes intersystem mobility policy data to the UE.
5. Sometime later, the UE requests by itself an access network discovery data; in order to avoid download of an unnecessary amount of data, and to optimize the usage, the UE may include its location information.
6. The ANDSF responds by a (filtered) set of data, focused for UEs current location.
7. The UE may switch on or off its non-3GPP radio interfaces, according to the information received in step 6, optimizing the scanning and battery consumption.
8. The UE can decide on access network selection and resulting handover.

The FQDN for ANDSF is constructed according to 3GPP's general DNS principles, with a prefix "andsf" and including MCC and MNC; an example is "andsf.mnc676.mcc232.pub.3gppnetwork.org".

5.13 SRVCC (Single Radio Voice Call Continuity) Procedure

Based on the architecture described in sub-section 3.5.2 and overall concept in sub-section 4.14.2, the message flows given in Figures 5-34 and 5-35 result for the Dual Transfer mode (i.e. the case where the UE has parallel radio capability). Note: for brevity, in the following the term "MSC server" should be understood as the "MSC server especially enhanced for SRVCC". The flow is simplified regarding the exact timing of messages, but it is in principle a maximum case: the target MSC is kept different from the MSC Server (a further simplification would result if they were identical); also the case with a non-voice component of bearers is shown. The complexity of the message flow demonstrates that an extensive effort was spent for the smooth interworking between CS and voice over PS domains. Where unique CS and PS parts of the message flow appear, they are marked accordingly in the graph.

As a prerequisite, the UE has to indicate in the attach procedure that it has SRVCC capability (this is indicated by a bit in UE's network capability), and the MME must be SRVCC-capable; also, SRVCC must be configured in the subscription data and CS related identifiers required in the interworking are to be sent to the MME and stored there.

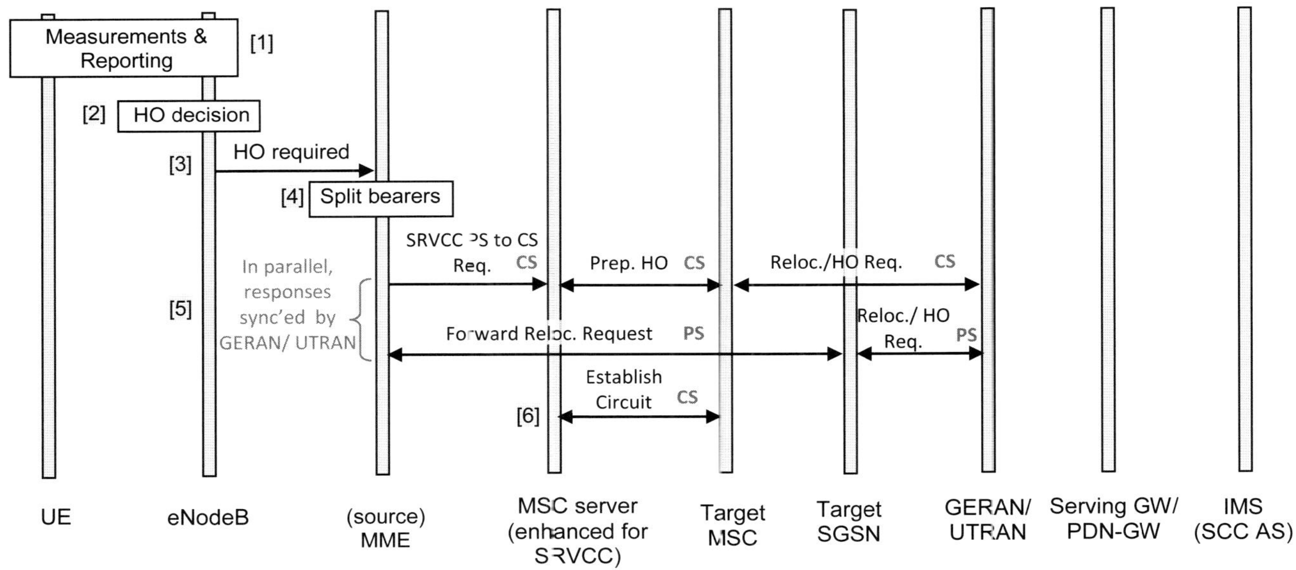

Figure 5-34: message flow for SRVCC from E-UTRAN to GERAN/UTRAN (part 1)

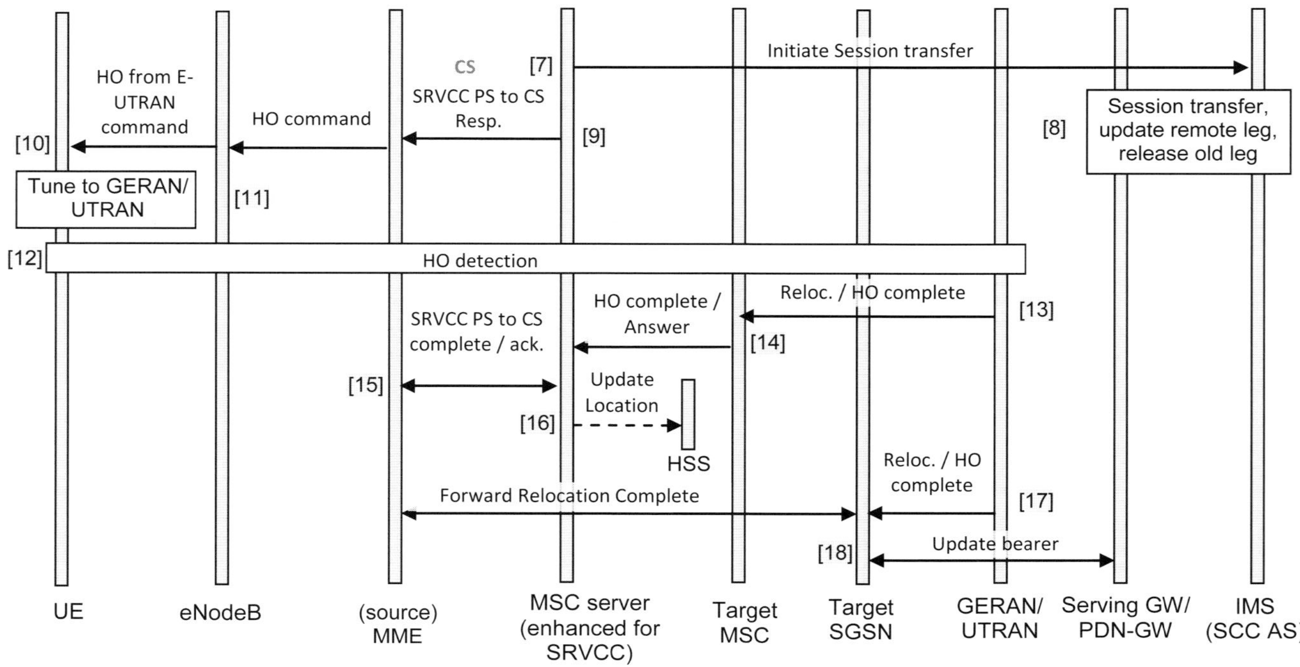

Figure 5-35 message flow for SRVCC from E-UTRAN to GERAN/UTRAN (part 2):

The steps in the somewhat complex message flow are explained as follows:

1. UE, based on instructions from the eNodeB, performs radio measurements of its current E-UTRAN cell and neighbouring GERAN/UTRAN cells; these are reported back to the eNodeB.
2. The eNodeB decides that a SRVCC handover should happen. This decision relies on the information that the UE supports SRVCC and that MME also supports it.
3. The eNodeB sends a "HO required" message to the MME, including RAN specific information in a transparent container and indicating SRVCC.
4. The MME splits the bearer with QCI=1 (voice bearer) from the rest of bearers.
5. MME initiates the PS-CS handover for the voice bearer and sends a corresponding request to the MSC server, from where this kind of request is further relayed to the target MSC by ISUP signaling (Prepare Handover); the MSISDN and a Session Transfer Number (STN) are included, for later reference. The target MSC in turn requests resources from the target GERAN/UTRAN by Relocation/Handover request. These messages are acknowledged back to the requesting entities. In parallel, the MME requests the relocation of PS resources from the target SGSN for the non-voice bearers, and the (target) SGSN in turn requests resources from the GERAN/UTRAN by virtue of a "Relocation/Handover request". Again, these messages are acknowledged back. The target GERAN/UTRAN synchronizes the message flow, as it waits for the requests for both requests, before sending the responses.
6. After having received the response from the CS target MSC, messages are exchanged to establish the traffic circuit (via ISUP signaling).
7. The MSC server initiates the session transfer by sending an ISUP IAM message using the Session Transfer Number; here the standard Service Continuity procedures are employed.
8. The IMS SCC Application Server continues the handling by transferring the session, updating the remote leg with SDP conforming to CS voice bearer and releasing the IMS (source) leg via E-UTRAN. From this point onwards the downlink IP (for voice) packets flow via the IMS/CS interconnecting node and are transformed to CS data streams.
9. The MSC server sends the response for PS to CS handover to the MME, including a transparent container of data (from the target system to the source system).
10. The MME synchronizes the two preparing message flow portions (for CS and PS) and sends a "HO command" to the eNodeB when both, corresponding acknowledgements have arrived. The instruction to actually execute the handover is given to the UE by a "HO from E-UTRAN command".
11. The UE tunes the radio channels to the (target) GERAN/UTRAN system.

12. The handover detection is performed (signaling from UE to the GERAN/UTRAN radio nodes and further on to the target MSC); for simplicity the details of this step are not shown.
13. The (target) GERAN/UTRAN notifies the completion of CS handover/relocation to the target MSC.
14. Because target MSC is different from the MSC server, the indication of completion needs to be forwarded on ISUP level, switching the speech circuit and including 'Answer' message.
15. The MSC server exchanges the specific completion messages pair with the originally requesting MME.
16. At this point in some cases the HSS needs to be queried by the MSC server (for correct operation of Supplementary Services and routing of terminating calls).
17. In parallel to steps 13 to 16, the completion of PS handover/relocation is executed; the target GERAN/UTRAN sends a corresponding message to the SGSN, where it is forwarded to the MME with acknowledgement back to SGSN.
18. The target SGSN updates the bearers status with the Serving GW, excluding the bearer formerly used for voice (over IMS PS).

5.14 IMS based Service Continuity

Detailing the concept of access transfer as presented in sub-section 4.17, the message flow is given in Figure 5-36; it involves the UE, IMS core nodes and the SCC application server (AS).

1. The UE has an ongoing IM session via IP-CAN 1; due to prior registration for the SC feature it is anchored in the SCC AS. Media paths are in place (for simplicity one per direction assumed).
2. The UE connects to IP-CAN 2, using another IP interface; it receives a second IP address.
3. The UE performs (a second) registration with the IMS (including SCC AS) via IP-CAN 2.
4. The UE reserves the same or appropriate amount of resources in IP-CAN, for media data and signaling; e.g. the codec(s) for media are most likely kept.
5. The UE sends a SIP INVITE with a new SDP offer to the IMS, indicating that this new session should overwrite the existing one and including the new contact address. At this point the UE is also ready to receive media (e.g. RTP) data over IP-CAN 2.

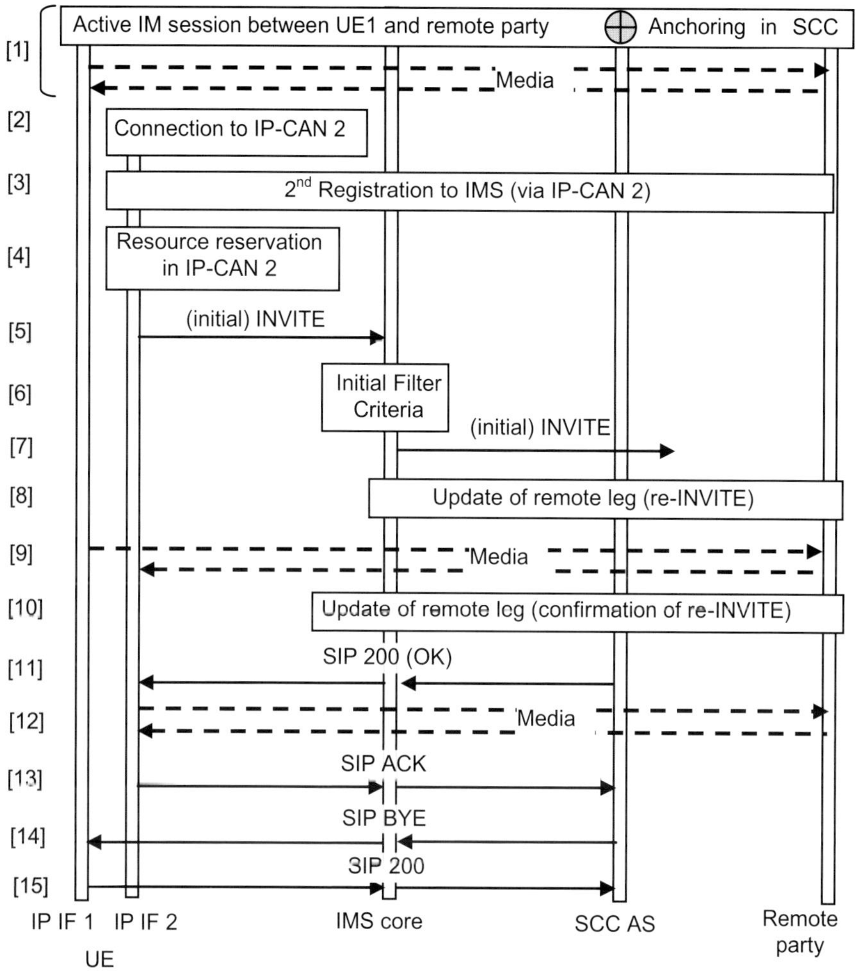

Figure 5-36: message flow for access transfer with IMS based Service Continuity

6. The initial INVITE is analyzed by the IMS and recognized as an originating new SIP request. Initial filter criteria for the UE are applied.
7. Based on the initial filter criteria, the INVITE is forwarded to the SCC AS.
8. The SCC AS correlates the incoming INVITE with the existing session to the remote party. It sends a re-INVITE to the remote party, including the new contact header (new IP address) and the (identical) SDP offer. This message is routed via the local IMS core (S-CSCF), where appropriate replacement of headers is done.

9. The remote party does not have to change any media properties, only the destination of its sent data packets. Thus, the downlink media path is updated immediately to be via IP-CAN 2, while uplink media is still routed via IP-CAN 1.
10. The re-INVITE is confirmed by SIP 200(OK) from the remote party to SCC AS via IMS core nodes and a SIP ACK in the reverse direction.
11. The initial INVITE of steps 6 and 7 is confirmed by SIP 200(OK) from the SCC AS via IMS core nodes and IP-CAN 2 to the UE.
12. From now on the UE sends also uplink media over IP-CAN 2.
13. The UE acknowledges finally the session dialogue for the initial INVITE of step 6 back to SCC AS.
14. The SCC AS sends a SIP BYE message to the UE via IP-CAN 1, to let release this access leg also for signaling.
15. The UE acknowledges back to the SCC via the IMS core. This is the final activity via IP-CAN 1.

5.15 Warning message delivery

The feature of warning messages was driven by earthquake and Tsunami disasters. As one can easily imagine, delivery of warning messages is a most critical task; e.g. if distribution is to incorrect destinations or fails in affected areas, the population would either panic unnecessarily and ignore subsequent warnings, or would remain without any warning even though it counted on it.

For delivery in E-UTRAN, the warning message flow spreads out downlink along the chain of nodes CBC – MMEs – eNodeBs – UEs, based on the architecture presented in sub-section 3.11 (only one instance of the latter three entities is shown in Figure 5-37). In order to notice any of the warning, the UE of course needs to be ETWS capable.

The warning messages originate at the Cell Broadcast Entity (CBE) and travel via Cell Broadcast Center (CBC) to the MMEs for distribution. As a precondition, the UE may be provisioned (via the OMA DM administrative "channel") with a list of networks from which it can accept warning messages without having bootstrapped a security relationship (note that in this context the important part of mutual security is authentication of the network to the UE). By default, this list is empty.

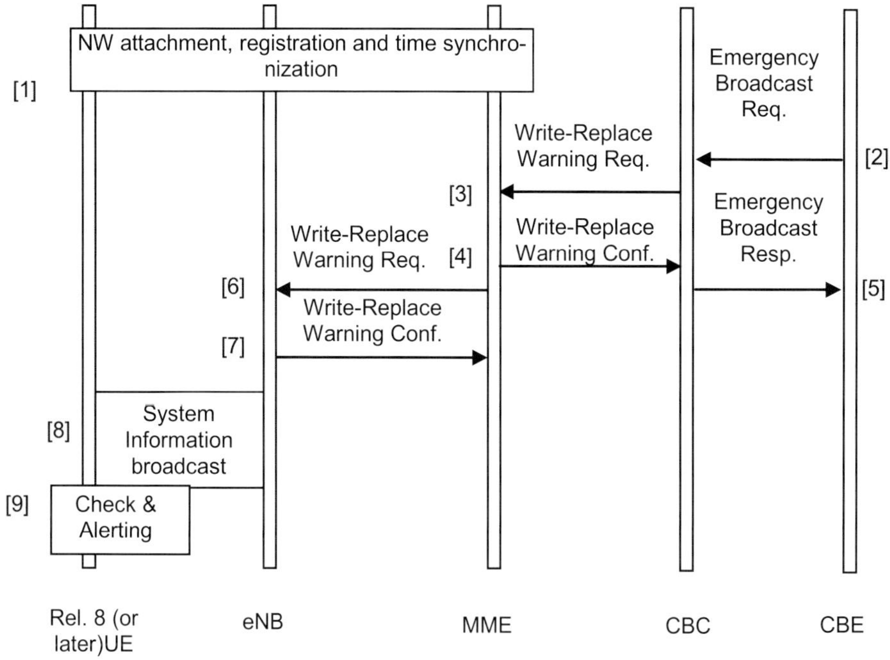

Figure 5-37: warning message delivery in E-UTRAN

The warning message flow in detail consists of these steps:

1. The UE(s) are attached and registered in the network; they synchronize their clock to the network.
2. The CBE sends the request for warning distribution to the CBC, including parameters for warning type, the warning message, the affected area and time period for the warning. (The concrete protocol on this interface is not defined by 3GPP.)
3. The CBC determines from the area information the target MMEs and sends the write/replace warning message containing a message identifier, serial number of the warning, tracking area ID (TAI) list, warning area and OMC ID).
4. The MME confirms back to the CBC that it has started the distribution of the warning.
5. The CBC confirms back to the CBE that it has started the distribution of the warning. (Note that in parallel the CBC also may distribute warnings to GERAN and UTRAN domains and report success thereof).
6. The MME forwards the warning notification to target eNodeBs. The TAI list is used to determine these (if empty, the warning needs to be distributed to all eNodeBs connected to the MME).

7. The eNodeB confirms the reception of the warning message back to the MME.
8. The eNodeB may receive duplicate warning messages, if it has multiple connections to MMEs, but they can be filtered out by analyzing message identifier and serial number. The eNodeB determines the cell(s) in which it needs to broadcast the warning (if one is already ongoing, it will be replaced) and starts broadcast of primary and secondary warning information.
9. The UE reads System Information and detects the primary warning notification. If security check for warnings is mandatory in this network, it checks the digital signature and timestamp. If present and the check succeeds, it reads the secondary notification and alerts the user accordingly; if not, it ignores warning notifications for the repetition period (this period is determined by regulation, or by default 60 seconds). If security check is not mandatory in this network, the UE alerts the user immediately, if the type of the warning is "earthquake" or "tsunami" or the combination of both; in the other case, the UE will check the digital signature and timestamp and, if the check is passed, read secondary warning and alert the user accordingly. The UE filters out duplicate warnings due to cell changes.
10. The eNodeB sends an indication of success or failure of the warning broadcast back to the originating MMEs, where a trace record is written.

The warning delivery in UTRAN is similar, with corresponding messages on the different path CBC-RNC-NB, from where a cell broadcast channel is used.

5.16 Procedures for I-WLAN mobility

5.16.1 PDN attach

As a prerequisite for this procdure, the UE has gained connectivity on the GPRS or I-WLAN level, based on the architecture given in sub-section 3.8. If now I-WLAN mobility is needed, the explicit attach to a PDN with mobility support and authorization is required. The information flows is seen in Figure 5-38 (simplified).

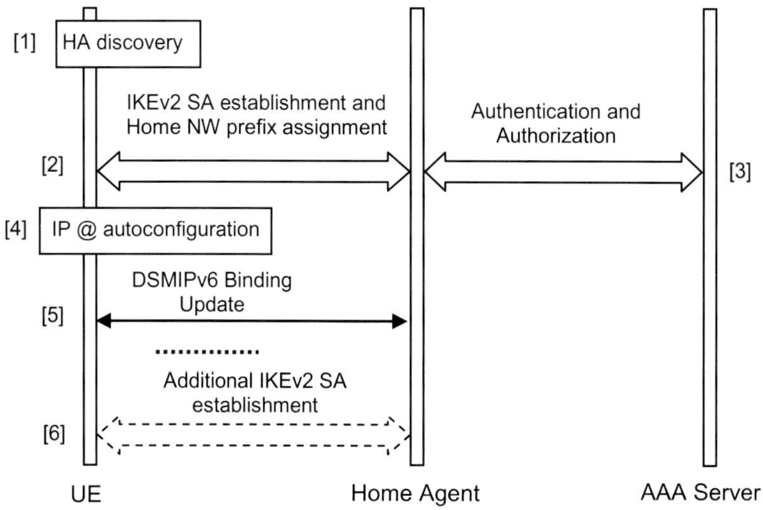

Figure 5-38: PDN attach procedure for I-WLAN mobility

1. The UE discovers the Home Agent (HA), by one of the methods in sub-section 5.2.6.
2. The UE initiates IKEv2 signaling towards the selected HA and establishes a Security Association (for protection of subsequent DSMIPv6 signaling) with it; it can optionally indicate a specific APN, for which it wants to access a PDN (otherwise the default APN would apply). The HA assigns an IPv6 Home NW prefix and relays EAP messages (via IKEv2) towards the UE.
3. The HA requests from the AAA server (via the H2 interface) the authentication of the UE and authorization of the mobility service, based on EAP (encapsulated in DIAMETER).
4. The UE performs IPv6 address autoconfiguration, i.e. generates from the assigned Home NW prefix a home address. The UE also detects if it is on the home link by comparing the received Home NW prefix with the prefix information received in Router Advertisements.
5. If the UE is not on its home link, it exchanges a DSMIPv6 binding update with the HA. The HA verifies the authenticity of the incoming message. The UE may request also an IPv4 home address, in which case the HA will include it in the response (Binding Update Acknowledgement) message.
6. At some later point (even though it could be also immediately) an additional Security Association can be established between UE and HA, for protection of user traffic; this is done by additional IKEv2 signaling.

The PDN attach procedure has to be repeated for every PDN the UE wants to be connected (i.e. effectively per APN).

5.16.2 UE initiated PDN detach

The procedure, sketched in Figure 5-39, is most simple and consists in sending a DSMIPv6 Binding Update message with lifetoime set to zero from the UE to the HA, and subsequent teardown of the Security Association by IKEv2 signaling between these two entities; in course of the latter, the HA communicates with the backend AAA infrastructure via H2 interface to clean up the I-WLAN mobility session.

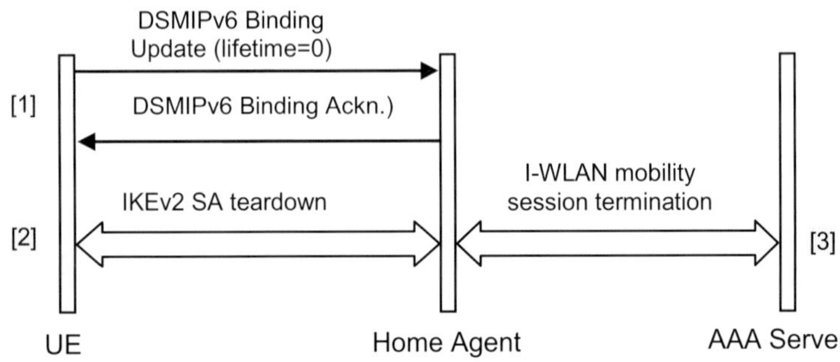

Figure 5-39: UE triggered PDN detach procedure for I-WLAN mobility

Two other variants of PDN detach exist (but are not shown here): AAA server triggered detach and HA triggered detach; both are network triggered, and thus cannot rely on standard (original) DSMIPv6 signaling. Instead, the enhanced signaling with "Binding Revocation" from the HA to the UE, currently being developed in IETF, is used.

5.16.3 Handover

Figure 5-40 shows a typical sequence for mobility between I-WLAN access and GPRS access (neglecting AAA server and HSS, and assuming for simplicity only one APN/GGSN/PDG in use). Here it is not assumed that one of the two mentioned accesses is the home link, in other words, the HA is separate from GGSN and PDG and not specially coordinated with them.

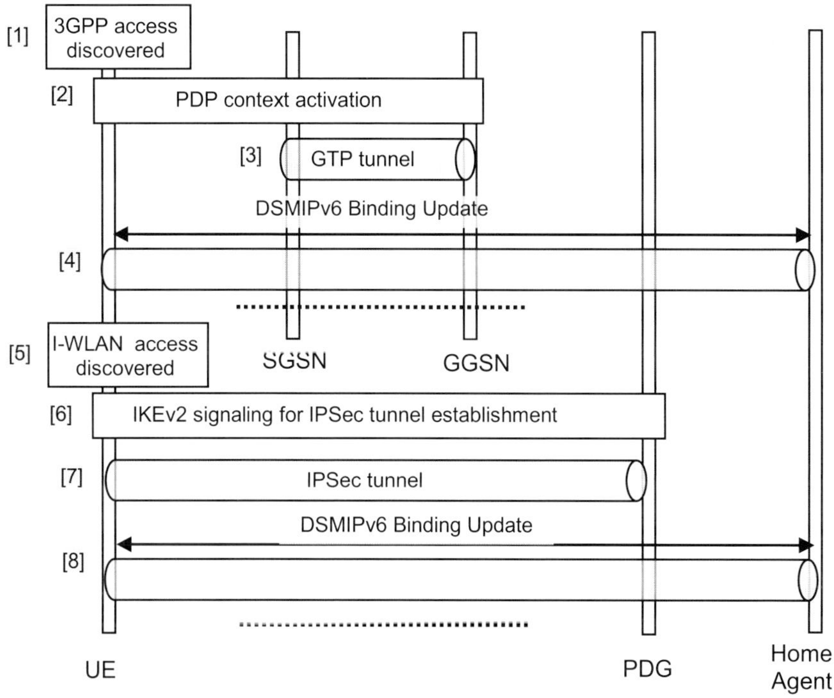

Figure 5-40: handover procedure for I-WLAN mobility

1. The UE, while being connected via I-WLAN, discovers 3GPP access and decides to use it with I-WLAN mobility, and thus shift all sessions (PDN connections) to it (DSMIPv6 based handover).
2. The UE sets up GPRS connectivity, by standard GPRS related signaling; a GGSN is selected, a PDP context is created, based on the APN in use, and an IP address assigned.
3. As a result, a GTP tunnel is established between SGSN and GGSN.
4. The IP address received in step 2 is used as a CoA in the DSMIPv6 Binding Update signaling, with the indication that home address preservation is requested, for the home address in use. (It is assumed that PDN attach has been performed before, so that the HA is already selected, a security association and IPSec tunnel established with the HA, a home NW prefix allocated and a home address in use, as explained in sub-section 4.15.3.) Now a DSMIPv6 tunnel is established or updated between UE and HA.
5. At some later point in time the UE discovers that I-WLAN access and decides to use it; this means a DSMIPv6 based handover is required.

6. The UE performs standard I-WLAN procedures for PDG selection, based on the APN in use. The UE then performs IKEv2 signaling for establishment of an IPSec tunnel between UE and PDG (this is also similar to the IPSec tunnel establishment between UE and ePDG as it occurs in case of access to EPC from untrusted non-3GPP access, see sub-section 5.4.4). An IP address is assigned by the PDG.
7. After successful completion of step 6, the IPSec tunnel is established between UE and PDG.
8. The UE exchanges a DSMIPv6 binding update with the HA, using the previously received I-WLAN specific IP address as CoA. Again, the preservation of home address is requested, and the DSMIPv6 tunnel is updated accordingly.

Note that partially two tunnels are in use: the DSMIPv6 tunnel is inside the IPsec tunnel (of I-WLAN access) or GTP tunnel (of GPRS tunnel).

References

[1] IETF RFC 3748 (June 2004): "Extensible Authentication Protocol (EAP)"
[2] IETF RFC 5448 (May 2009): "Improved Extensible Authentication Protocol Method for 3rd Generation Authentication and Key Agreement (EAP-AKA')"
[3] IETF RFC 4555, "IKEv2 Mobility and Multihoming Protocol (MOBIKE)"
[4] IETF RFC 2131 (March 1997): "Dynamic Host Configuration Protocol"
[5] 3GPP TS 33.222: "Generic Authentication Architecture (GAA); Access to network application functions using Hypertext Transfer Protocol over Transport Layer Security (HTTPS)"

Chapter 6: Protocol Environment

Protocols actually implement the signaling and data exchange within a mobile network. As has been shown in preceding chapters, for the new system many new interfaces had to be introduced or enhanced. In order to carry all necessary information flows across them in a reliable and efficient manner, 3GPP's protocol designers were challenged with a huge amount of work.

In this chapter we walk through the protocols of major importance, and for each of them explain these basics:

- general concept and usage,
- protocol structure,
- protocol message structure,
- overview of supported protocol messages, plus one or two illustrative examples, and
- if felt useful, single, selected message parameters.

It is by far not feasible and also not desirable to aim for completeness with respect to the last two items in this list. The interested reader should instead consult the cited sources of information.

6.1 Evolution of GPRS Tunneling Protocol (GTP)

6.1.1 General

GTP is the protocol developed for tunneling and encapsulation of data units and control messages in GPRS. Since its design in the late 1990s it was put into deployment on the large scale, and massive experience has been gathered. It was practically not possible, and also not desirable to give it up for the new system, but on the other hand it was immediately clear that enhancements were also needed, in order to be able to interwork smoothly with the legacy PS world and to support the functions required for the new system itself.

GTP for the Evolved 3GPP system comes in two variants, control and user plane. The control plane GTP-C handles the signaling, and it is needed in addition to the protocol for pure transfer of user related data, GTP-U; the latter is termed "user plane". Current versions, suitable for EPS, are GTPv1-U and GTPv2-C. (Note that another variant, GTP', is used for transfer of charging data records in the legacy system, but is not described here.)

☞ GTPv1-U and GTPv2-C are documented in 3GPP TS 29.281 [1] and 3GPP TS 29.274 [2], respectively.

The distinctive feature of GTP is that it supports the separation of traffic into "bearers" inside its main GTP tunnel; or in other words, the ability to bundle bearers and handle them together. This is on contrast to e.g. PMIP (see below in sub-section 6.2), and it has quite drastic consequences with respect to policy and charging control. In Figure 6-1 the tunneling concept of GTP and necessary means for identification are visualized.

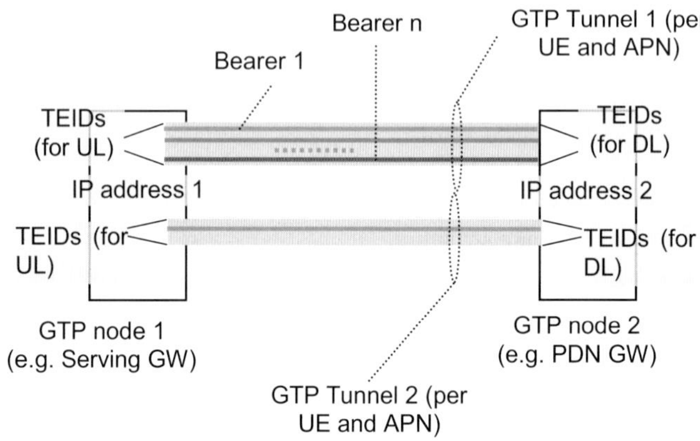

Figure 6-1: GTP tunneling concept

The endpoints of GTP tunnels are identified by TEIDs (Tunnel Endpoint Identifiers); they are assigned locally for uplink and downlink by the peer entities and signaled crosswise between them. Only both address items together, the TEID and the IP address of the GTP peer identify uniquely a data stream (additionally the UDP port number needs to be considered).TEIDs are used on different granularity, e.g. specific per PDN connection on S5 and S8 and per UE on S3/S4/S10/S11 interfaces.

6.1.2 GTPv2-C (Control Plane of GPRS Tunneling Protocol Version 2)

This protocol is used on the EPC signaling interfaces (including SGSNs of at least Rel. 8), e.g.:

- S3 (between SGSN and MME),
- S4 (between SGSN and Serving GW),
- S5 and S8 (between Serving GW and PDN GW; note that an alternative is offered by PMIPv6, see sub-sections 4.15.2 and 6.2),
- S10 (between two MMEs), and
- S11 (between MME and Serving GW).

For SRVCC GTPv2-C is also used between MME/SGSN and MSC (Sv interface). For interworking with CDMA2000 GTPv2-C is also used on S101 (between MME and an HRPD access network).

The protocol stack of GTPv2-C appears most simple, and is shown in Figure 6-2. The IP layer may be IPv4 or IPv6.

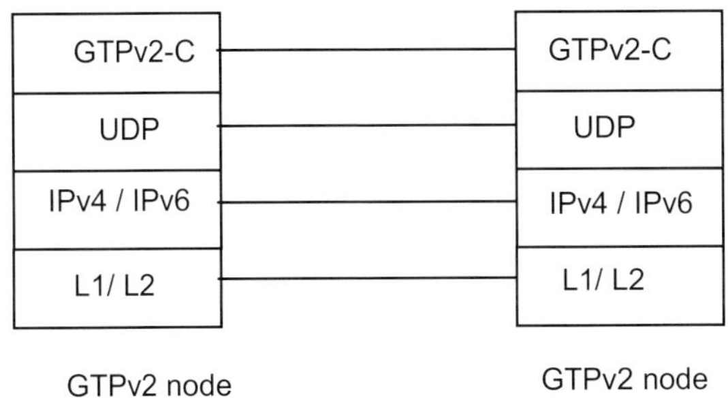

Figure 6-2: protocol stack for GTPv2-C

Corresponding to that, a typical GTPv2-C protocol data unit looks like shown in Figure 6-3: the GTP specific part is preceded by IP and UDP headers, it consists of a GTPv2-C header and a part containing GTPv2-C information elements in varying number, length and format, depending on the type of the message.

There are variations, e.g. for control messages like "Echo" and the notification that a protocol version is not supported, the TEID information is not present (this is indicated by the "T" flag set to "1"). The version is obviously firmly set to "2" in this protocol version.

Legacy GTP had a somewhat complex extension header mechanism; it is not used any longer in GTPv2-C.

The message type is set in the second octet (so at maximum 256 messages can be defined for future extensions). Table 6-1 gives an overview of the currently defined GTPv2-C messages. The length of the message is coded in octets 3 and 4 (counted in octets and excluding the first 4 octets themselves).

TEID is the tunnel endpoint identifier, a unique value at the remote/receiving side; it allows to multiplex and demultiplex tunnels at an endpoint, in the very frequent case that more than one GTP tunnel needs to be distinguished.

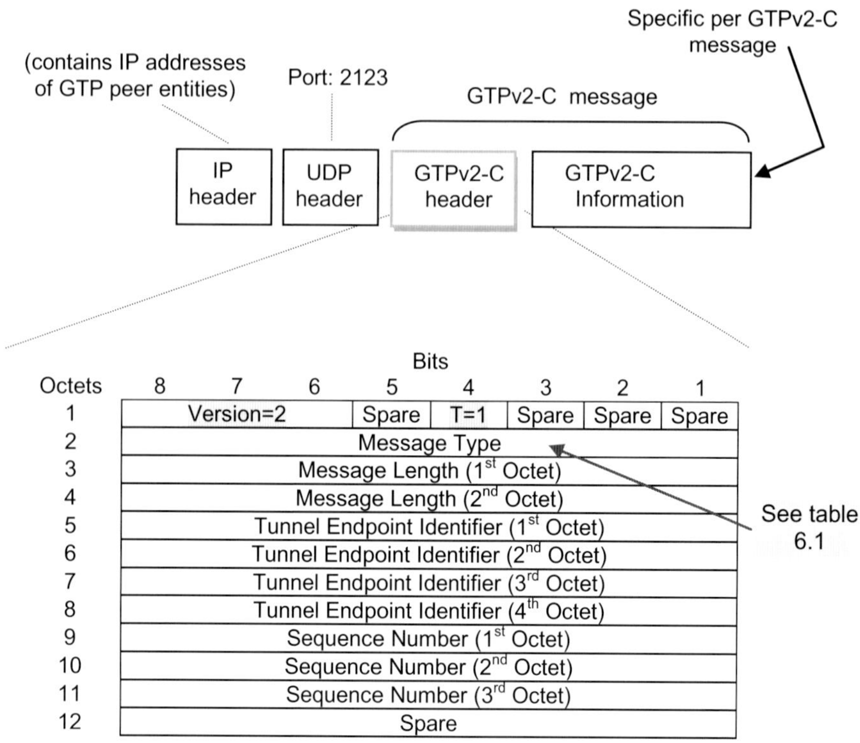

Figure 6-3: protocol data unit for GTPv2-C (typical header format including TEID)

Message type	Message	Additional Explanation
0	Reserved	Shall never be used (intentionally excluded from protocol, to enforce explicit setting).
1/2	Echo Request / Response	Used to probe if a GTP peer entity is alive.
3	Version Not Supported Indication	Contains the latest GTP version supported by the sending node.
4/5	Direct Transfer Request / Response	Used for tunneling signaling messages on S101 interface for optimized handover, between HRPD access node and MME and vice versa (see sub-section 5.10.4).
6/7	Notification Request / Response	Used for tunneling notifications (e.g. handover complete) on S101 between HRPD access node and MME and vice versa (see sub-section 5.10.4).

Message type	Message	Additional Explanation
25/26	SRVCC PS to CS Request / Res- ponse	Used to trigger and confirm SRVCC initiation between SGSN/MME and MSC server (see sub-section 5.13).
27/28	SRVCC PS to CS Complete Noti- fication / Acknowledge	Used to indicate and confirm completion of SRVCC between MSC server and SGSN/MME (see sub-section 5.13).
32/33	Create Session Request / Response	Used to establish connectivity between two nodes (e.g. in initial attach, when establishing additional PDN connectivity, when relocating a node).
34/35	Modify Bearer Request / Response	Used to modify properties of a single or of mul- tiple bearers; includes bearer context informa- tion.
36/37	Delete Session Request / Response	Tears down GTP control sessions.
38/39	Change Notification Request / Res- ponse	Used for reporting location information.
66 / 67	Delete Bearer Command / Delete Bearer Failure Indication	Instructs nodes to delete bearers and confirms back.
68 / 69	Bearer Resource Command / Bear- er Res. Failure Indication	Used to allocate or modify resources (ons S11, S5/8 and S4).
73	Stop Paging Indication	Sent from SGW to the MME or SGSN on S11/S4.
95/96	Create Bearer Request / Response	Instructs nodes to install bearer(s) and confirms back.
97/98	Update Bearer Request / Response	Used to inform the control plane nodes from the user plane about bearer changes.

Table 6-1: examples of message types for GTPv2-C (incomplete list)

It is not feasible and also not desirable to list here information elements for all, or even a major portion of GTPv2-C messages. But for illustration the contents of the “Create Session Request” (a tunnel management message) is presented in Table 6-2. Column “P” denotes the presence requirement of an information element (M … mandatory, O … optional, C … conditional).

IE	P	Explanation
IMSI	M	Unique permanent identification of the UE.
MSISDN	C	Mirrored from legacy ("telephone number").
ME Identity (MEI)	C	Unique identity of the Mobile Equipment (terminal, device).
User Location Info	C	Different formats are supported. For legacy accesses: Cell Global Id, Service Area Id and Routing Area Id; for E-UTRAN access: Tracking Area Id and E-UTRAN Cell Global Identifier.
Serving Network	C	Identity of the chosen network, i.e. MCC + MNC.
RAT Type	M	The type of access the UE is using (currently defined values are UTRAN, GERAN, WLAN, GAN, HSPA Evolution, EUTRAN).
Indication Flags	M	Encodes more details of conditions in the network (see explan. below).
Sender F-TEID	M	Fully qualified TEID for the control plane (sent to the other peer for use in its response messages).
PGW S5/S8 Address	C	PDN GW related address, sent e.g. from MME to Serving GW (for control plane / GTP or PMIP).
Access Point Name (APN)	C	For determining the PDN GW and point of interconnection with a PDN, to be set in initial attach and UE requested PDN connectivity procedures.
Selection Mode	C	Indicates whether the APN was the subscribed one or selected by MME. Nitial attach and UE requested PDN connectivity.
PDN Type	M	Indicates target IP version (IPv4, IPv6 or IPv4v6 = dual stack), based on HSS subscription and UE request.
PDN Address Allocation	C	This is the address information allocated by the network (in initial attach or UE requested PDN connectivityContains e. Either IPv4 address or IPv6 prefix, or both.
APN-AMBR	C	See sub-section 4.11, can be included for initial attach and UE requested PDN connectivity procedures.
Protocol Config. Options (PCO)	O	Carries detailed information on server/node addresses, see sub-section 4.10
Bearer Contexts to be created	M	List of bearers; per bearer e.g. the following items can be included: bearer id, TFT, addresses (TEIDs), QoS.
Trace Information	C	Includes structured information for tracing (like triggering event, IP address of encollecting entity).
Recovery	C	For recovery purposes. Included if the otherGTP peer is contacted for the first time (after restart).
MME / Serving GW CSID	O	Connection set identifier (for recovery purposes).
UE timezone	O	Offset between universal time and local time (where UE is currently roaming); given in steps of 15 minutes.
Charging characteristics	C	Defines a profile and details of the charging method to be applied (online or offline charging).
Private Extension	O	Generic means to extend the protocol; contains an "enterprise id".

Table 6-2: information elements for "Create Session Request" message (incomplete list)

The indication flags contain e.g. such information:

- S5/S8 Protocol Indicator: to indicate GTP or PMIP based S5/S8;
- Dual Address Bearer Flag: indicates if dual addressing is generally supported in SGSNs);
- Handover Indication: indicates if the UE is handing over from non-3GPP access;
- Operation Indication: codes whether a Serving GW relocation is involved in TAU/RAU or X2 handover;
- Direct Tunnel Flag: whether Direct Tunnel (from RNC to GGSN or Serving GW) is used;

6.1.3 Enhanced GTPv1-U

Only a small, but effective enhancement was applied for GTP-U, and for that purpose it was not deemed necessary to step up the protocol version number. Thus, we still look at GTPv1-U, but at least at its newest Rel. 8.

The protocol stack is practically the same as for GTPv2-C (see Figure 6-4), with only the name of layers and protocols substituted accordingly. The extension header mechanism (already in use before rel. 8) is kept in place; it allows to insert, when needed, two additional, originally not foreseen information elements:

1. UDP source port of the triggering message (two octets);
2. PDCP PDU number: related to the feature of lossless handovers; in this case data packets need to be numbered within the EPC (two octets).

The enhancement is the capability to transmit an "end marker" in the user plane. It is used during the inter-eNodeB handover procedure and gives the indication that the path is switched immediately after this data packet (such functionality was not required in pre-Rel.8, because GTP-U did not terminate in the radio access node (i.e. not in the BS or NodeB).

Only a few messages exist in GTPv1-U, and they are listed in Table 6-3. It is visible that in fact some very reduced kind of "signaling" is possible via GTPv1-U (echo and end marker mechanisms). The only message transferring "real" user data is type 255, the so-called G-PDU message; the only information element it carries after the header is the original data packet from a UE or external PDN.

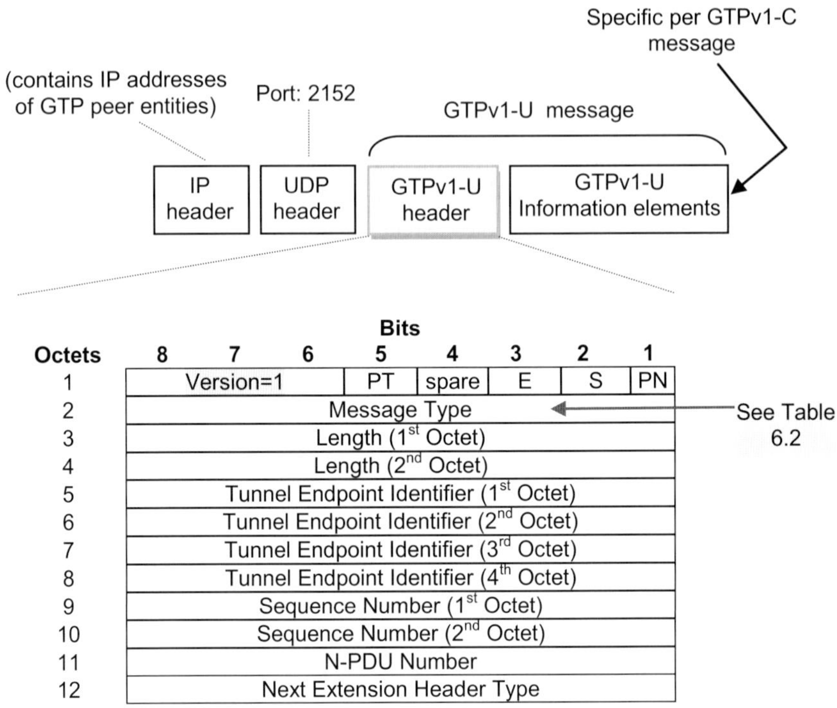

Figure 6-4: protocol data unit for GTPv1-U (typical header)

Message Type value	**GTPv1-U Message**
0	Reserved
1 /2	Echo Request / Response
3-25	Reserved (partially for GTPv1-C and GTP')
26	Error Indication
27-30	Reserved (for GTPv1-C)
31	Supported Extension Headers Notification
32-253	Reserved (partially for GTPv1-C and GTP')
254	End Marker
255	G-PDU

Table 6-3: GTPv1-U messages

Not all occurrences of GTP-U tunnels are listed in the reference architecture (which was intended to capture only longer lived associations between network nodes); temporary tunnels are possible:

- between two Serving GWs: applicable for the S1-based handover, in the case that the Serving GW is relocated. Due to the short-lived connection no path maintenance is required;
- between two SGSNs: corresponds to the previous case, but in the legacy PS network;
- between two RNCs: applicable for the relocation of RNCs in the 3G PS network (no relation with EPC, it is mentioned here just for completeness).

6.2 PMIPv6 protocol details

6.2.1 Dual Stack Capability

Dual stack capability for PMIPv6 has two targets:

1. to support IPv4 home addresses and
2. to allow IPv4 only transport across the access network; in this case the MAG may use also an IPv4 private address, and a NAT may be deployed along the path towards the LMA.

These two features can be used independently. To solve these requirements, the following extensions have to be made, as visualized in Figure 6-5 for two UEs (one with both an IPv4 Home Address and an IPv6 Home network prefix, the other one with currently only an IPv4 Home Address assigned):

- in the Binding Cache of LMA:
    - IPv4 home address assigned to the mobile node's interface and now registered by the mobile access gateway (includes the corresponding subnet mask). It comes either from static configuration/profile, or is dynamically allocated by LMA.
    - IPv4 default-router address assigned to the mobile node.
- In the Binding Update list of MAG:
    - IPv4 home address assigned to the mobile node's attached interface (includes the corresponding subnet mask).
    - IPv4 default-router address of the mobile node (received from LMA through the Proxy Binding Acknowledgment message).

The LMA and MAG need to implement IPv6, and they also need an IPv4 address. MAG is the IPv4 default-router for the UE on its access link.

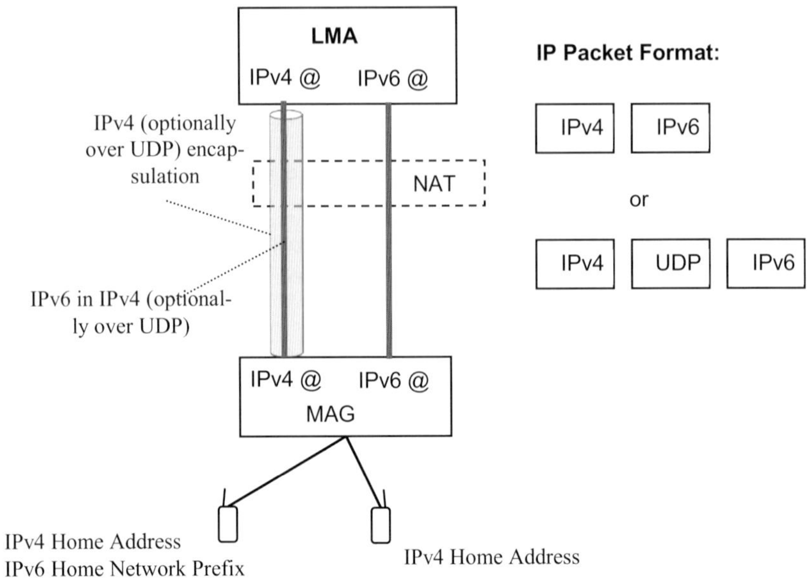

Figure 6-5: basic concept for PMIPv6 with IPv4 support

In the signaling messages therefore two new options are present:

(a) IPv4 Default-Router Address,

(b) IPv4 DHCP Support Mode.

The latter describes if the MAG should act as DHCP relay or DHCP server. The IPv4 address allocation and related signaling for the two variants is shown in Figure 6-6. The signaling messages exchanged between MAG and LMA are according to PMIPv6, but encapsulated in IPv4 or in IPv4 with UDP headers (the latter provides NAT support); the same happens with the user traffic.

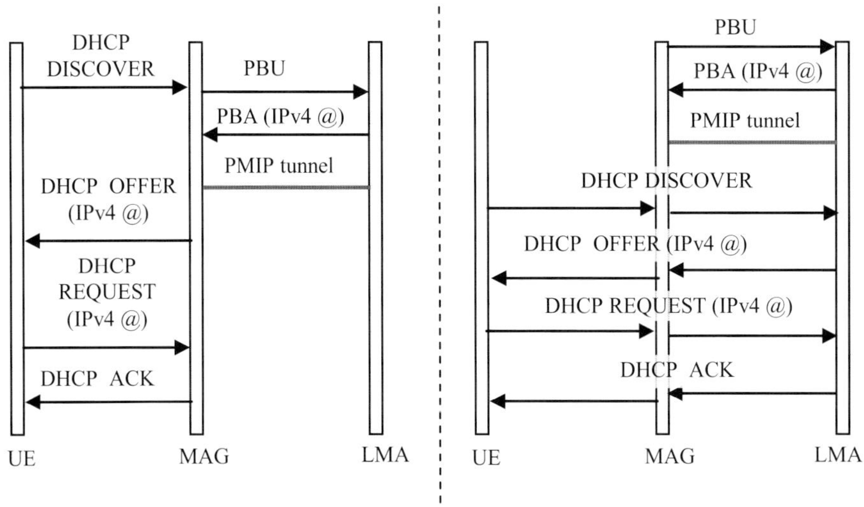

Figure 6-6: IP address allocation related signaling for MAG acting as DHCP server (left) and as DHCP relay (right)

6.2.2 PMIPv6 Signaling

Table 6-4 gives an overview of PMIPv6 signaling messages (PMIPv6 base signaling plus enhancements specified in IETF for Binding Revocation and path management). The basic PMIPv6 signaling is done with "Binding Update" (BU) from MAG to LMA, and a corresponding "Binding Update Acknowledgement" (BUA) messages back to MAG, this is used for binding registration, binding refreshment and binding deletion.

IP address information (IPv4 address or IPv6 address prefix) is normally requested by the LMA and allocated by the MAG in initial registrations.

PMIPv6 Signaling Message		Direction	Description
Binding Update	PBU	MAG → LMA	Requests creation, extension and deletion of a mobility binding; it is also used to request a new IPv4 address (in conjunction with DHCPv4 signaling of the UE).
Binding Update Acknowledgement	PBA	LMA → MAG	Acknowledges (positively or negatively) the requests for creation, extension and deletion of a mobility binding; it is also used to allocate an IPv4 address (in conjunction with DHCPv4 signaling of the UE).
Binding Revocation Indication	BRI	LMA → MAG	Notification that a binding is revoked and thus will be deleted by LMA; allows also bulk revocations.
Binding Revocation Acknowledgement	BRA	MAG → LMA	Acknowledges (positively or negatively) a binding revocation.
Heartbeat	HB	MAG → LMA, LMA→ MAG	Periodic signaling message, used for detecting failures.

Table 6-4: PMIPv6 messages

PMIPv6, as described so far, cannot yet do the same type of handovers as designed by 3GPP. Currently the specification of the LMA does not allow to accept uplink packets delivered from a MAG in a handover target access network, before the PMIP signaling for update of the binding has taken place. Also, by this binding update the whole path is switched; it is thus not possible to have a triangular routing, so that downlink packets are still routed via the old access, while uplink user traffic would already go via the new MAG up to LMA (see Figure 6-7).

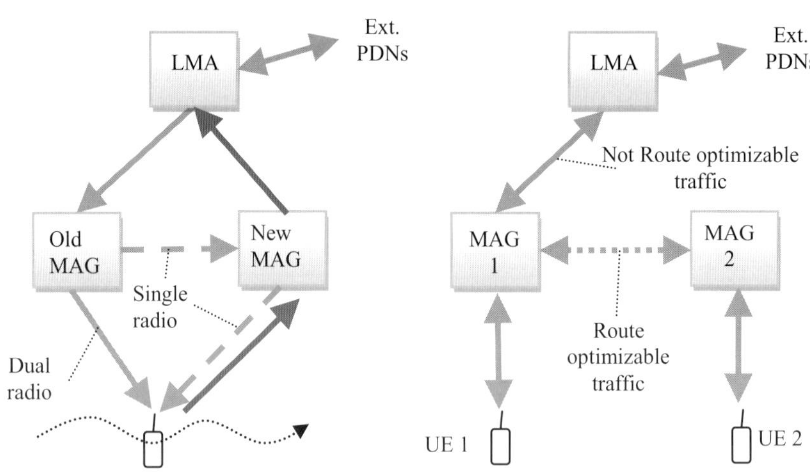

Figure 6-7: proposed triangular routing (left) and route optimization (right) with PMIP

Extensions are already proposed in the protocol IETF design group; the first one consists in transient binding, where the opening of the new uplink path does not switch the PMIP tunnel for downlink. IETF is further working on route optimization, which is not included in the original PMIPv6 protocol. By this, the traffic between two communicating UEs served by PMIP MAGs under the same LMA (e.g. one in PMIP based EPC, by Serving GW and another one in trusted non-3GPP access) could be shortcut. Still, these new ideas need consideration of security, charging and policy and corresponding enhancements in 3GPPs specification.

6.2.3 3GPP Specific Information Elements added to PMIPv6

PMIPv6 was designed for a very general usage; 3GPP had a few special requirements, stemming from the need to make it compatible as much as possible with GTP capabilities. Luckily, the base mobility protocol MIP already allows so-called "vendor specific extensions" ("vendor" in this case maps to 3GPP); they could be readily carried over to PMIPv6. 3GPP therefore could defined 5 extensions, as described in Table 6-5.

Vendor Specific Information	**Direction**	**Explanation**
Protocol Configuration Options	MAG→ LMA, LMA → MAG	Mirrored from GTP; used to transfer frequently needed, protocol related data between UE and network.
Specific 3GPP related error code	LMA→ MAG	E.g. it can indicate that no access is given to an APN, or that the GRE key is missing in a signaling message.
Connection Set Identifier (CSI) list	LMA→ MAG, MAG→ LMA	Contains one or more CSIs. It is generated for each new PDN connection, and used in case of partial node failure to identify the PDN connections.
PDN type indication	LMA → MAG	Used to indicate the decision of the PDN GW, if a change in PDN type (IPv4 or IPv6) had to be applied, plus a cause associated with it.
PDN GW IP address	MAG→LMA	Used in case of chaining on S2a/S2b to transfer to the intermediate LMA (Serving GW) the address of the higher LMA (PDN GW).
DHCPv4 address allocation indication	LMA → MAG	Indicates that IP @ allocation via DHCPv4 is to be used by the UE (thus no IPv4 address to be allocated in the initial PBA).

Table 6-5: Vendor Specific options in PMIPv6 messages (when used by 3GPP)

For the transport of APN information the "Service Selection" option, already defined in IETF ([5], [6]) is used.

6.3 Extensible Authentication Protocol (EAP)

Extensible Authentication Protocol (EAP) is a generic framework developed by IETF (RFC 3748). The base signaling mechanism supports different authentication "methods" on top.

The specific usage of EAP for interworking with a 3GPP system is defined by the EAP-AKA "method" (note that the abbreviation "AKA" may seem general, but it specifically means "3GPP Authentication and Key Agreement", and NOT a general one). EAP-AKA is already used in I-WLAN.

The concept is like shown in Figure 6-8. A functional entity, the authenticator, is placed in the access network and communicates with the device subject for authentication on L2 on the front side, and a special node, the Authentication Server, anywhere in the backend of the network infrastructure. Specific variants of EAP are defined for the different L2 technologies, e.g. over LAN and WLAN links, but also higher layer protocols acting as "virtual layer 2" support it, like IKEv2 signaling. The backend signaling is realized typically over AAA type of protocols, e.g. RADIUS and DIAMETER (the Evolved 3GPP system only uses the latter one, but RADIUS is still an option for EAP transport within the 3GPP legacy system). Note: the described procedure conforms to the so-called pass-through mode of an authenticator, which is exclusively used in the 3GPP context.

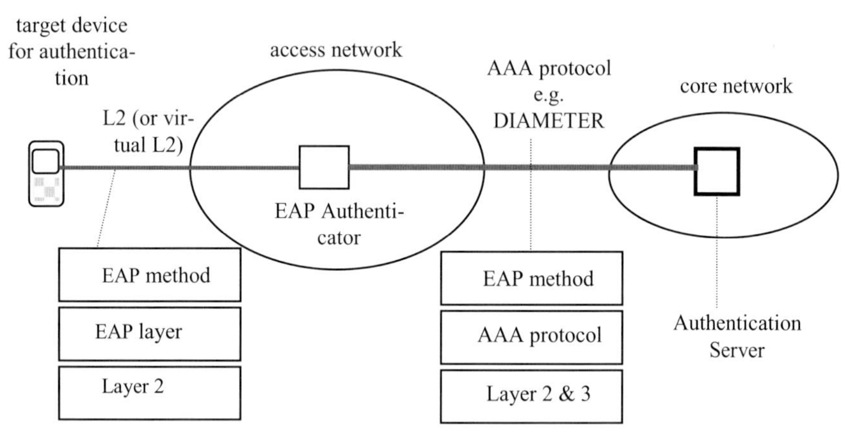

Figure 6-8: EAP concept

The principal steps for EAP authentication are these (see also IETF RFC 4187 [3]):

1. the EAP authenticator issues an authentication request towards the target device/UE (on L2); it receives the response from the target device/UE and passes it on to the AAA infrastructure;
2. the AAA server runs the EAP method, resulting in a challenge for the target device, which is sent via the authenticator;
3. the target device has to answer the challenge; the respond is relayed back to the AAA server via the authenticator;
4. the AAA server compares the response to the challenge with the expected one and decides on the success of authentication. Either a success or failure indication is sent back to the target device.

Optionally, notifications can be used to transfer additional information; this is utilized for IP Mobility Mode Selection indication (see sub-section 4.15.6).

Durin the design there was a principal decision to separate the security domains of non-3GPP access networks from the 3GPP security domain, and also from each other. The reasoning is that in this way a security breach in one of these domains does not endanger the security within the 3GPP system (and other non-3GPP access networks). The practical consequence is that the ID of the non-3GPP access network enters the security algorithm, which requires the specification of a variant of EAP-AKA, the EAP-AKA' (prime), see IETF RFC 5448 [4].

6.4 Internet Key Exchange Protocol Version 2 (IKEv2) and MOBIKE

Internet Key Exchange is a sophisticated framework defined in version 2 by IETF in RFC 4306 [7]. It allows to set up and maintain Security Associations and IPSec tunnels between two nodes and to exchange some configuration data; these are transferred in so-called configuration payloads in the message dialogues. A full IKEv2 session consists of several dialogues, structured into phases.The typical and basic message flow is given in Figure 6-9, together with a description of how it is applied in the context of signaling between UE and ePDG:

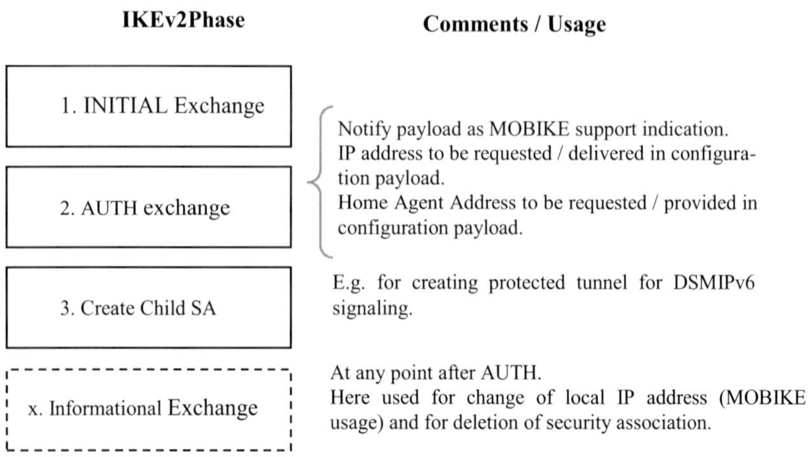

Figure 6-9: basic procedure for IKEv2

In the Evolved 3GPP system IKEv2 is used for:

a) Establishing the IPSec tunnel required for access from an untrusted non-3GPP network: the endpoint in the network is ePDG. The configuration data exchanged are:
   - IP address information: either IPv4 address or IPv6 prefix.
   - IP mobility mode selection information:

b) Establishing a secure signaling channel for DSMIPv6: the endpoint in the network is either the PDN GW (in case the full EPC architecture is used), or the HA (in case of I-WLAN). The configuration data exchanged are:
   - IP address information: IPv6 prefix.
   - Optionally, a DNS server address.

MOBIKE (defined in RFC 4555) extends the IKEv2 protocol with mobility and multihoming; in practice it means that the UE may change its local IP address and update the other node. In EPC it is used for realizing the mobility between untrusted non-3GPP access networks, if they connect to the same ePDG (see sub-section 5.10.3). In this case, the support for MOBIKE must be signaled beforehand, in a notification part of the initial IKEv2 signaling.

6.5 DIAMETER

DIAMETER is a generic AAA protocol, with added functions for network access, mobility and QoS handling. Although in principle of a general peer-to-peer nature, it is used in 3GPP's architecture in client-server mode. It has a built-in extensibility, and thus ideally supports message structures on interfaces with the need for some flexibility. Also, it supports multiple server configurations with failure and failover handling. Functionally it has similarities with its predecessor RADIUS (which is also used on various interfaces of 3GPP's legacy architecture), but differs profoundly on the message and parameters level. DIAMETER provides a dead peer detection capability by heartbeat message pairs. It may be run over SCTP or TCP and uses port number 3868.

The 'vendor' identification for 3GPP's DIAMETER applications is quite naturally "3GPP" (10415).

The DIAMETER protocol is used extensively in the EPC, e.g. on:

- S6a (between MME and HSS): for subscription download and update.
- S6d, which is the counterpart of S6a for the legacy world with interworking capability with the new system (between an upgraded SGSN and HSS): for subscription download and update.
- S13 (between MME and EIR): for equipment checking.
- SWa (between untrusted non-3GPP access and AAA server): for authentication.
- STa (between trusted non-3GPP access and AAA server): for authentication and authorization.
- SWd (between an AAA proxy and a AAA server): for forwarding between VPLMN and HPLMN.
- S6b (between PDN GW and AAA server): for authorization (of APN and mobility).
- SWm (between ePDG and AAA server): for authentication and authorization.
- SWx (between AAA server and HSS): for exchange of authentication vectors and registration information.
- Gx (between PDN GW and PCRF), Gxa (trusted non-3GPP access network and PCRF) and Gxc (Serving GW and PCRF): for IP-CAN session handling and GW-Control Session handling.

The applications used are those for Credit Control (Request and Re-Authorization), EAP payload conveyance and MIPv4 / MIPv6 support.

The EAP related information remains completely transparent to the DIAMETER signaling; the information needed for the specif key derivation of EAP-AKA' (prime) is not included in the EAP payload, transported in a separate parameter (Access Network ID).

The following message pairs and triples are used, according to the flow principle:

- Diameter-EAP-Request (DER) / Diameter-EAP-Answer (DEA) on STa: for initial authentication and authorization from trusted non-3GPP access;
- Abort-Session-Request (ASR) / Abort-Session-Answer (ASA) on STa: for network initiated detach;
- Session-Termination-Request (STR) / Session-Termination-Answer (STA) on the STa interface : for termination of a session by the trusted non-3GPP access network;
- Re-Auth-Request (RAR) / AA-Request (AAR) / AA-Answer (AAA) on STa: for handling updates of information, triggered by the network.

The content of authentication/authorization request message via STa and SWm is given in Table 6-6 (column "P" codes the presence requirement).

Message parameter	Sta	SWm	P	Explanation
User Identity	X	X	M	NAI form of identification is used (according to IETF RFC 4282).
EAP payload	X	X	M	Encapsulated EAP payload; used for mutual authentication between UE and AAA server.
Authentic. Req. Type	X	X	M	Defines whether user is to be authenticated only, authorized only or both.
UE Layer-2 address	X		M	Link layer address of the UE.
Supported 3GPP QoS profile			O	May be sent if the non-3GPP access network supports QoS mechanisms (relies on recent enhancements of DIAMETER).
Mobility Capabilities	X		C	Indicates PMIPv6 and DHCPv6 support by non-3GPP access (relevant for dynamic mobility mode selection).
Mobility Features		X	C	Indicates PMIPv6 support (relevant for dynamic mobility mode selection) and IKEv2 based Home Agent address discovery by the ePDG.
Access Type	X	X	Sta: M; SWm: O	Contains non-3GPP access network technology type.
Access NW Identity		X	M	Contains the access network identifier used for key derivation of EAP-AKA' at the HSS.
Visited NW Identifier	X	X	O	Identiy inserted by the non-3GPP access network.
APN	X	X	M	Stems from UE request (but transport between UE and non-3GPP access network is access technology dependent, and between UE and ePDG it is optional).
Terminal Information	X		O	Contains information about the user's mobile equipment (e.g. used if non-3GPP access network technology is HRPD).

Table 6-6: parameters of authentication request messages via STa and SWm

For the optional 3GPP based access authentication in untrusted non-3GPP access message contents is largely the same, but exclude service related information, and access network related information is optional.

6.6 Stream Control Transmission Protocol (SCTP)

SCTP is a reliable transport protocol operating on top of a potentially unreliable connectionless packet service such as IP. It was developed especially for signaling application and offers acknowledged, error-free non-duplicated transfer of datagrams (messages). Detection of data corruption, loss of data and duplication of data is achieved by using checksums and sequence numbers. A selective retransmission mechanism is applied to correct loss or corruption of data. The decisive difference to TCP is multihoming and the concept of several streams within a connection. Where in TCP a stream is referred to as a sequence of bytes, an SCTP stream represents a sequence of messages (and these may be very short or long). SCTP tries to combine advantages of UDP and TCP, but avoid their disadvantages; it is defined in IETF RFC 4960 [8].

The concept of SCTP is shown in Figure 6-10:

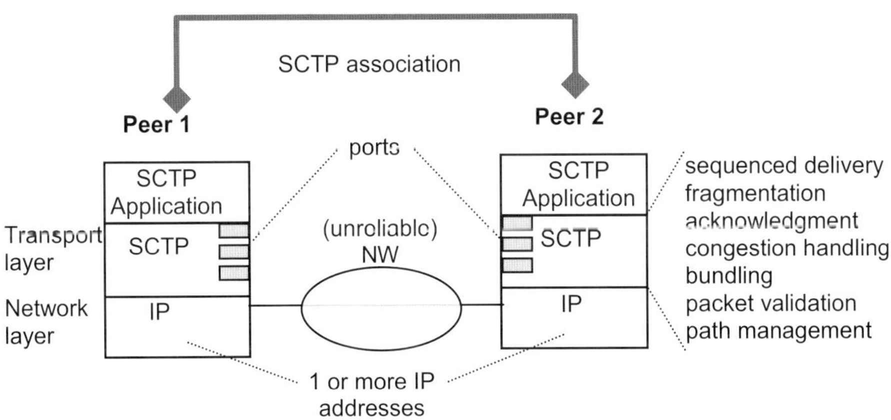

Figure 6-10: SCTP concept

SCTP is used on several network internal control plane interfaces, with these SCTP applications:

- S1-MME (between eNodeB and MME): S1-AP for AS-NAS signaling;
- SBc (between MME and SBc): SBc-AP for warning message delivery; IANA registered port number is 29168.
- S6a (between MME and HSS) and S6d (between SGSN and HSS): a specific DIAMETER application is used on top of SCTP, for subscriber data download and registration; the application id assigned by IANA is 16777251.
- SGs (between MSC/VLR and MME): SGs-AP for coordinative signaling between CS domain and E-UTRAN for the purpose of CS fallback; port number is 29118.
- S13 (between MME and EIR): a specific DIAMETER application (application id assigned by IANA is 16777252) is used on top of SCTP, for checking of equipment numbers.

6.7 S1 Application Protocol

Two categories of procedures across S1-MME exist: UE associated and non-UE associated. Additionally two classes of messages are defined: class1 is with and class 2 is without a response. Class 1 procedures and related initiating/response messages are listed in Table 6-7; for class 2 procedures the message names are largely identical to the procedure names, thus Table 6-8 lists only these.

Elementary Procedure	Initiating Message	Response (if successful)
Handover Preparation	HANDOVER REQUIRED	HANDOVER COMMAND
Handover Resource Allocation	HANDOVER REQ	HANDOVER REQ ACK
Path Switch Request	PATH SWITCH REQ	PATH SWITCH REQ ACK
Handover Cancellation	HANDOVER CANCEL	HANDOVER CANCEL ACK
E-RAB Setup	E-RAB SETUP REQ	E-RAB SETUP RESP
E-RAB Modify	E-RAB MODIFY REQ	E-RAB MODIFY RESP
E-RAB Release	E-RAB RELEASE COMMAND	E-RAB RELEASE RESP
Initial Context Setup	INITIAL CONTEXT SETUP REQ	INITIAL CONTEXT SETUP RESP
Reset	RESET	RESET ACK
S1 Setup	S1 SETUP REQ	S1 SETUP RESP
UE Context Release	UE CONTEXT RELEASE COMMAND	UE CONTEXT RELEASE COMPLETE

UE Context Modification	UE CONTEXT MODIFICATION REQ	UE CONTEXT MODIFICATION RESP
eNodeB Configuration Update	ENB CONFIGURATION UPDATE	ENB CONFIGURATION UPDATE ACK
MME Configuration Update	MME CONFIGURATION UPDATE	MME CONFIGURATION UPDATE ACK
Write-Replace Warning	WRITE-REPLACE WARNING REQ	WRITE-REPLACE WARNING RESP

Table 6-7: Class 1 S1-AP procedures and related messages

Elementary Procedure	**Elementary Procedure**
Handover Notification	Error Indication
E-RAB Release Indication	UE Context Release Request
Paging	Downlink S1 CDMA2000 Tunneling
Initial UE Message	Uplink S1 CDMA2000 Tunneling
Downlink NAS Transport	UE Capability Info Indication
Uplink NAS Transport	eNodeB Status Transfer
NAS non delivery indication	MME Status Transfer
Deactivate Trace	Overload Start
Trace Start	Overload Stop
Trace Failure Indication	eNodeB Direct Information Transfer
Location Reporting Control	MME Direct Information Transfer
Location Reporting Failure Indication	eNodeB Configuration Transfer
Location Report	MME Configuration Transfer
Cell Traffic Trace	

Table 6-8: Class 2 S1-AP procedures

The parameters contained in S1-AP messages and their semantics for eNodeB vary greatly. As an example, for the messages “Initial UE message”, “Uplink NAS Transport” and “Downlink NAS Transport” the essential information element is the NAS protocol data unit. It is handled like a container, and eNodeB does no interpretation. Additionally, the “Initial UE message” carries mandatorily the Tracking Area identity and the E-UTRAN cell identity from which the NAS message was sent by the UE (note that it might differ from the cell id corresponding to the eNodeB which just now handles the Sa-AP message, due to ongoing handover), CSG identity to which the eNodeB belongs (in case of being an eNodeB) and an establishment cause. The latter two are missing in the “Uplink NAS Transport” message, because a signaling context is already in place; therefore in this message also two related identifiers,”MME UE S1AP ID” and eNodeB “UE S1AP ID” are included. Clearly “Initial UE message” has only “UE S1AP ID”, as it is just about to initialize the signaling relation between eNodeB and MME. On the other hand, "Downlink NAS transport" may include a handover restriction list.

Altogether approximately 70 radio network related, 3 transport layer related, 3 SON related and 27 NAS layer related information elements have been defined in the S1-AP protocol.

6.8 X2 Application Protocol

X2 application protocol has much in common with the S1-AP; the same categorization into class1 and class 2 messages is made. The message set is much smaller, corresponding to the specialized function of X2 (see Table 6-9 for an overview).

Procedure	Initiating Message	Class	Response (if successful)
Handover Preparation	HANDOVER REQUEST	1	HANDOVER REQUEST ACKNOWLEDGE
Reset	RESET REQUEST	1	RESET RESPONSE
X2 Setup	X2 SETUP REQUEST	1	X2 SETUP RESPONSE
eNodeB Configuration Update	ENB CONFIGURATION UPDATE	1	ENB CONFIGURATION UPDATE ACKNOWLEDGE
Resource Status Reporting Initiation	RESOURCE STATUS REQUEST	1	RESOURCE STATUS RESPONSE
Load Indication	LOAD INFORMATION	2	
Handover Cancel	HANDOVER CANCEL	2	
SN Status Transfer	SN STATUS TRANSFER	2	
UE Context Release	UE CONTEXT RELEASE	2	
Resource Status Reporting	RESOURCE STATUS UPDATE	2	
Error Indication	ERROR INDICATION	2	

Table 6-9: X2-AP procedures

6.9 NAS Signaling Protocol

6.9.1 General

The NAS signaling protocol is genuine 3GPP development, thus nowhere else than in 3GPP systems found. It defines:

- the possible states in the UE for mobility and session management;
- the counterpart in MME (for EPC) and legacy network nodes (SGSN for GPRS and MSC for CS domain);
- messages exchanged between both sides;
- the conditions/context in terms of procedures (for the state models see subsection 4.9 and 4.10).

In the following we concentrate on the protocol stack variant for EPC, and only occasionally refer to similarities or equivalences with the legacy variants.

The NAS signaling protocol is in principle security protected; integrity protection is applied where/whenver possible. Yet, in the early phases of signaling, when the security context is not yet bootstrapped or available as stored from earlier attachments, some NAS signaling messages need to be sent unprotected.

6.9.2 NAS signaling protocol for EPS mobility management

NAS signaling procedures for EPS mobility management are listed in table 6-10. The "X" in column "C" indicates if a variant for the combination with the corresponding CS related NAS signaling protocol stack exists (this is the case for a configuration with CSFB enabled, by virtue of the so-called "combined" procedures). Only the success cases are shown (error cases are produced by using appropriate reject messages, mostly by the network); the explanation also does not include all possible cases.

The message dialogues generally are guarded by timers against loss of messages; e.g. timer T3410 is used to supervise the attach procedure and expires after 15 seconds. If then no response (acceptance or rejection) from the network has been received, a retry procedure is started. Counters are used to limit the retrys. For EPS mobility nmanagement the network runs 7 timers and the UE 14 timers.

Procedure	C	Messages	Explanation
GUTI reallocation		GUTI REALLOCATION COMMAND ←	Used to allocate a temporary identifier (GUTI) to the UE (to support privacy).
		GUTI REALLOCATION COMPLETE →	
Authentication		AUTHENTICATION REQUEST ←	Used to authenticate the UE (i.e. check whether its identity is the one it claims to be).
		AUTHENTICATION RESPONSE →	
Security mode control		SECURITY MODE COMMAND ←	Used to negotiate between UE and MME about security algorithms and to take a security context into use.
		SECURITY MODE COMPLETE →	

Identification		IDENTITY REQUEST ← IDENTITY RESPONSE →	Used to determine a UE's identity, if not known/derivable from the information received before.
EMM information		EMM INFORMATION ←	Used to transfer support information like network name and time zone from the network to the UE.
EMM status		EMM STATUS ← or →	Used for error reporting.
Attach	X	ATTACH REQUEST → ATTACH ACCEPT ← ATTACH COMPLETE →	Used to register the UE with the network and to allocate default resources.
Detach	X	DETACH REQUEST → or ← (DETACH ACCEPT ← or →)	Used to remove a UE's registration with the network and to de-allocate all resources.
Tracking area updating	X	TRACKING AREA UPDATE REQUEST → TRACKING AREA UPDATE ACCEPT ←	Used for idle mode mobility (if the UE enters cells belonging to different Tas than those with which is is registered). It makes the network aware of UE's location on the TA granularity, and to synchronize information between UE and network.
Service Request		SERVICE REQUEST → or EXTENDED SERVICE REQUEST →	E.g. used when the UE is in idle mode and has UL data to send, and also as a response to paging (see next).
Paging		Request to lower layer for paging SERVICE REQUEST →	Used when UE is in idle mode and downlink traffic arrives. The "request to lower layer" is a message to all eNodeBs in all currently assigned Tas (a "cross-layer" request).
Transport of NAS message		UL NAS TRANSPORT → or DL NAS TRANSPORT ←	Used for SMS transport in encapsulated form to/from MSC/VLR (as necessary in case of CSFB).

Table 6-10: NAS signaling procedures and messages for EPS mobility management

6.9.3 NAS signaling protocol for EPS session management

The second block of NAS signaling functionality is related to session handling; four network initiated and four UE initiated ESM procedures exist. Table 6-11 lists them with the corresponding messages flows for the successful case (again, negative cases are created by using appropriate reject messages). Some of them are piggybacked onto EMM NAS messages, where a container is provided (e.g. a

PDN CONNECTIVITY REQUEST is packed onto the ATTACH REQUEST message). Additionally two simple messages have been defined for general information exchange.

Procedure	Messages	Explanation
Default EPS bearer context activation	ACTIVATE DEFAULT EPS BEARER CONTEXT REQ. ←	Establishes a default EPS bearer context between the UE and the EPC; is performed in response to a UE requested PDN connectivity.
	ACTIVATE DEFAULT EPS BEARER CONTEXT REQ. →	
Dedicated EPS bearer context activation	ACTIVATE DEDICATED EPS BEARER CONTEXT REQUEST ←	Establishes a EPS bearer context with specific QoS and TFT between the UE and the EPC. It may be requested by the UE through a Bearer Resource Allocation procedure or initiated by the network.
	ACTIVATE DEDICATED EPS BEARER CONTEXT REQUEST →	
EPS bearer context modification	MODIFY EPS BEARER CONTEXT REQUEST ←	Modifies an existing EPS bearer context for specific QoS and TFT between the UE and the EPC. It may be requested by the UE through a Bearer Resource Allocation or Bearer Resource Modification procedure (see below), or initiated by the network.
	MODIFY EPS BEARER CONTEXT ACCEPT →	
EPS bearer context deactivation	DEACTIVATE EPS BEARER CONTEXT REQUEST ←	De-activates an EPS bearer context or disconnects from a PDN. It may be requested by the UE through a Bearer Resource Modification procedure or PDN Discconect or initiated by the network.
	DEACTIVATE EPS BEARER CONTEXT ACCEPT →	
UE requested PDN connectivity	PDN CONNECTIVITY REQUEST →	Requests set up of a default bearer to a PDN.
	ACTIVATE DEFAULT EPS BEARER CONTEXT REQ. ←	
UE requested PDN disconnect	PDN DISCONNECT REQUEST →	Disconnects UE from one PDN (while being connected to at least another one); all bearers for this PDN are deleted. Note: disconnect from all PDNs is achieved by a detach.
	DEACTIVATE EPS BEARER CONTEXT REQUEST ←	
UE requested bearer resource allocation	BEARER RESOURCE ALLOCATION REQ. →	Used to request bearer resources.
	MODIFY EPS BEARER CONTEXT REQ. ← or ACTIVATE DEDIC. EPS BEARER CONTEXT REQ. ←	
UE requested	BEARER RESOURCE MODIFICATION REQUEST →	Used to modify (includes also de-alocate) bearer resources.

bearer resource modification	ACTIVATE DEDIC. EPS BEARER CONTEXT REQUEST ← or MODIFY EPS BEARER CONTEXT REQ. ←	
ESM information request	ESM INFORMATION REQUEST ←	Used to request protocol configuration options or APN from the UE (in case they could not be included in the initial NAS message due to security requirements.
	ESM INFORMATION RESPONSE →	
ESM status	ESM STATUS → or ←	Informs the other NAS signaling peer of errors.

Table 6-11: NAS signaling procedures and messages for EPS session management

The UE initiated ESM messages are "UE requested PDN connectivity", "UE requested PDN disconnect", "UE requested bearer resource allocation" and "UE requested bearer resource modification"; the four network initiated ESM messages are "Default EPS bearer context activation", "Dedicated EPS bearer context activation", "EPS bearer context modification procedure" and "EPS bearer context deactivation".

6.9.4 Message coding and examples of information elements

The general message coding is given in Figure 6-11. The variable part varies greatly in length, from zero (e.g. in AUTHENTICATION REJECT or DETACH ACCEPT, where no specific information element needs to be transferred) up to hundreds or theoretically thousands of octets (e.g. in ATTACH REQUEST or ATTACH ACCEPT, due to their piggybacked ESM message containers).

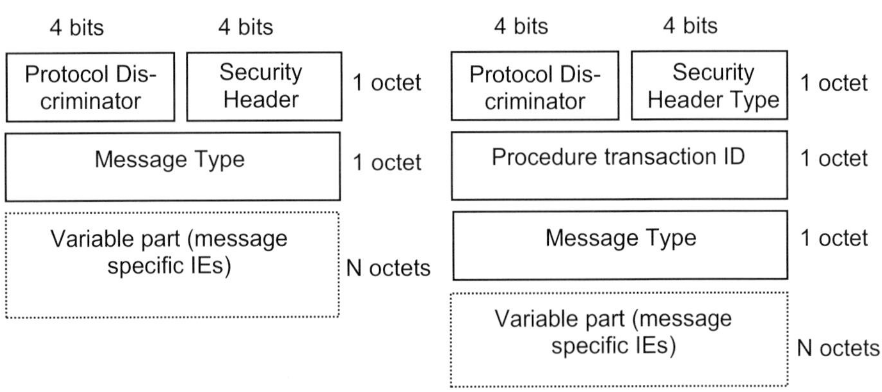

Figure 6-11: coding of EMM (left) and ESM (right) NAS signaling messages

Per NAS signaling message mandatory and optional Ies are defined; presence of the latter depends on further conditions and context.

NAS signaling messages sent as negative response (e.g. ATTACH REJECT or SECURITY MODE COMMAND REJECT) contain a cause, to instruct the UE for further actions. E.g.if the attach reject cause #12 (Tracking area not allowed) is received by the UE, it needs to delete its GUTI, last visited registered TAI and KSI, and also to store the current TAI in the list of "forbidden tracking areas for regional provision of service" and enter the state EMM-DEREGISTERED.LIMITED-SERVICE.

The procedure transaction ID in ESM messages is used to identify messages across transactions (i.e. several related messages sent between UE and network).

As an illustrative example, the specific Information Elements contained in BEARER RESOURCE MODIFICATION REQUEST and ATTACH ACCEPT messages are given in Table 6-12 and Table 6-13.

Information Element	Explanation	P	Length (octets)
Linked EPS bearer ID	Identifies default bearer	M	½
Spare half octet	To fill up the format (zero bits)	M	½
Traffic flow aggregate	Set of packet filters describing the service data flow (see sub-section 4.12.3)	M	2-256
Required traffic flow QoS	Set of the new target Quality of Service parameters pertaining to the bearer	M	3-10
Protocol configuration options	Included in the message when the UE wishes to send or request configuration parameters (see also sub-section 4.10)	O	3-253

Table 6-12: contents of BEARER RESOURCE ALLOCATION REQUEST message (only variable part)

Information Element	Explanation	P	Length (octets)
EPS attach result	Indicates if the attachment was done for EPS services only, or as a combined attach for EPS and CS services (for CSFB).	M	½
Spare half octet	Filler, set to zero	M	½
T3412 value	Value for UE's periodic tracking area update timer.	M	1
TAI list	Tracking Areas (max 16) assigned to the UE.	M	7-97
ESM message container	Contains the piggy-backed ESM message.	M	2-n
GUTI	Temporary (EPS) identifier assigned to the UE.	O	13

Location area identification	Used to assign a new location area from the CS domain, if the attach was a combined one.	O	6
MS identity	Used to assign a new temporary identifier from the CS domain, if the attach was a combined one.	O	7-10
EMM cause	Gives further information about the reason, if the ATTACH ACCEPT is not fully successful (i.e. if combined attach was requested but only attach procedure in EPS was successful).	O	2
T3402 value	Attach attempt related timer value	O	2
T3423 value	ISR related timer value	O	2
Equivalent PLMNs	List of PLMN which are handled equivalent to the HPLMN in PLMN selection.	O	5-47
Emergency Number List	Emergency number(s) for use within the current PLMN.	O	5-50

Table 6-13: message contents of ATTACH ACCEPT message (only variable part)

In one more step of detailing, we show the sub-structure of parameter "Required traffic flow QoS" in Table 6-14, although without going down to the bit coding level:

Field (octets)	**Explanation**
EPS quality of service IEI	Identifies this Information Element
Length of EPS quality of service contents	Codes the length of this Inform. Element
QCI	QoS Class Index (see sub-section 4.11)
Maximum bit rate for uplink	Detailed coding is not shown here; it allows a granularity of 1 kbit/s from 1 to 64 kb/s, granularity of 8 kbit/s from 64 kbit/s to 568 kbit/s and a granularity of 64 kbit/s from 576 kbit/s to 8640 kbit/s without extension fields. With extension fields the granularity is 100 kbit/s from 8700 kbit/s to 16000 kbit/s, 1 Mbits/s from 17 Mbit/s to 128 Mbit/s and 2 Mbit/s from 130 Mbit/s to 256 Mbit/s.
Maximum bit rate for downlink	
Guaranteed bit rate for uplink	
Guaranteed bit rate for downlink	
Maximum bit rate for uplink (extension)	
Maximum bit rate for downlink (extension)	
Guaranteed bit rate for uplink (extension)	
Guaranteed bit rate for downlink (extension)	

Table 6-14: coding of "Required traffic flow QoS" parameter

For non-GBR QCIs, the maximum and guaranteed bit rates are ignored.

6.10 OMA Device Management

Open Mobile Alliance (OMA) is a standardization body in charge of network independent service enabling functions. One of their 'enablers' is taking care of Device Management, in effect providing means (i.e. an information exchange protocol) to manage configuration data for and on devices. This is done by so called Managed Objects; they describe the structure, type and semantics of the various configurable data items.

For the Evolved 3GPP system the OMA DM scheme [9] is used for:

- Access network Discovery and Selection data (provided by ANDSF as the OMA DM server),
- Allowed CSG lists (provided by the CSG list server),
- Disabling/Enabling of features of the UE,
- IMS related configuration (including network operator's voice domain selection preferences).

3GPP has therefore defined these five Management Objects and registered them with OMA. They contain a threefold but corresponding description (graphical, textual and DDF/XML code); see as a simple examples of Managed Object definition for Allowed CSG lists in Figure 6-12 (data "tree") and Figure 6-13 (DDF code).

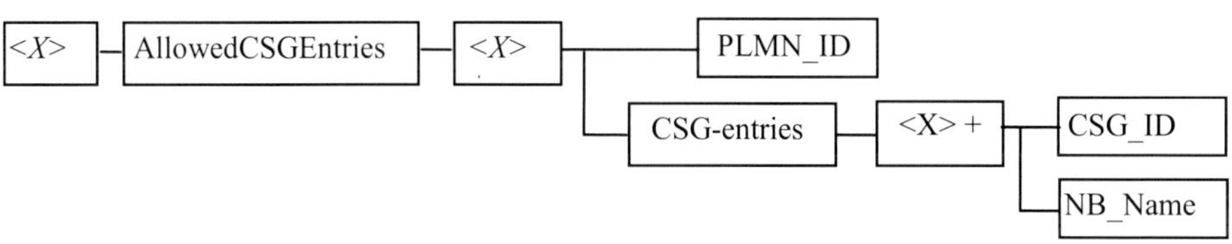

Figure 6-12: example of a data "tree" structure of a Managed Object in OMA DM (Allowed CSG list)

```
<?xml version="1.0" encoding="UTF-8"?>
<!DOCTYPE MgmtTree PUBLIC "-//OMA//DTD-DM-DDF 1.2//EN"
"http://www.openmobilealliance.org/tech/DTD/dm_ddf-v1_2.dtd">

<MgmtTree>
     <VerDTD>1.2</VerDTD>
     <Node>
          <NodeName/>
          <DFProperties>
               <AccessType> <Get/> </AccessType>
               <DFFormat> <node/> </DFFormat>
               <Occurrence> <ZeroOrOne/> </Occurrence>
               <DFTitle>Root Node of the UE Allowed CSG List</DFTitle>
               <DFType>
                    <DDFName>urn:oma:ext-3gpp-csg:1.0</DDFName>
               </DFType>
          </DFProperties>
          <Node>
               <NodeName>AllowedCSGEntries</NodeName>
               <DFProperties>
                    <AccessType> <Get/> </AccessType>
                    <DFFormat>  <node/> </DFFormat>
                    <Occurrence> <One/> </Occurrence>
                    <DFTitle>This node specifies the parent node for allowed
```

Figure 6-13: example of DDF coding of a Managed Object definition in OMA DM (Allowed CSG list; extract)

References

[1] 3GPP TS 29.281: “General Packet Radio System (GPRS) Tunnelling Protocol User Plane (GTPv1-U)”

[2] 3GPP TS 29.274: “Evolved Packet System (EPS); Evolved General Packet Radio Service (GPRS) Tunnelling Protocol for Control plane (GTPv2-C); Stage 3”

[3] IETF RFC 4187 (January 2006): “Extensible Authentication Protocol Method for 3rd Generation Authentication and Key Agreement (EAP-AKA)”

[4] IETF RFC 5448 (May 2009): “Improved Extensible Authentication Protocol Method for 3rd Generation Authentication and Key Agreement (EAP-AKA’)”

[5] IETF RFC 5149 (February 2008): “Service Selection for Mobile IPv6”

[6] IETF RFC 5446 (February 2009): “Service Selection for Mobile IPv4”

[7] IETF RFC 4306 (December 2005): “Internet Key Exchange (IKEv2) Protocol”

[8] IETF RFC 4960 (September 2008), “Stream Control Transmission Protocol”

[9] OMA-ERELD-DM-V1_2: “Enabler Release Definition for OMA Device Management”

Annex: Information stored in MME and Serving GW

The following two tables are included for illustration purposes only, not for reference. Data items IMSI (primary key) and MSISDN, well known from the legacy system, are not shown. Similarly, trace related data (trace reference, trace type, trigger id and OMC entity) are omitted for brevity.

Data stored in MME

Field	Description
MM State	Mobility management state ECM-IDLE, ECM-CONNECTED, EMM-DEREGISTERED.
GUTI	Globally Unique Temporary Identity of the UE.
ME Identity	Mobile Equipment Identity e.g. IMEI/IMEISV
Tracking Area List	Current Tracking Area list
TAI of last TAU	TAI of the TA in which the last Tracking Area Update was initiated.
EUTRAN Cell Global Identity	Last known E-UTRAN cell
E-UTRAN Cell Identity Age	Time elapsed since the last E-UTRAN Cell Global Identity was acquired
Authentication Vector	Temporary authentication and key agreement data; consists of four elements: a) network challenge RAND, b) expected response XRES, c) Key K_{ASME}, d) network authentication token AUTN.
UE Radio Access Capability	UE radio access capabilities.
MS Classmark 2	GERAN/UTRAN CS domain core network classmark (used if MS supports SRVCC to GERAN or UTRAN)
MS Classmark 3	GERAN CS domain radio network classmark (used if MS supports SRVCC to GERAN)
Supported Codecs	List of codecs supported in the CS domain (used if the MS supports SRVCC to GERAN or UTRAN)
UE Network Capability	UE network capabilities including security algorithms and key derivation functions supported by the UE
MS Network Capability	For a GERAN and/or UTRAN capable UE, this contains information needed by the SGSN
Selected NAS and AS Algorithms	Selected NAS and AS security algorithm
K_{ASME} , KSI_{ASME}	Main key for E-UTRAN key hierarchy based on CK, IK and Serving network identity, and the corresponding key set identifier
NAS Keys and COUNT	K_{NASint}, K_{NASenc}, and NAS COUNT parameter
E-UTRAN/UTRAN Key Set flag	Indicates whether UE is using security keys derived from UTRAN or E-UTRAN security association

Table A-1: data stored in MME (part 1)

Selected CN operator id	Selected core network operator identity (to support network sharing).
Recovery	Indicateor for HSS performing DB recovery.
Access Restriction	Access restriction subscription information.
ODB for PS parameters	Status of operator determined barring for PS services.
MME IP@, TEID for S11	MME IP address and tunnel endpoint identifier for S11 interface (used by Serving GW)
Serving GW IP@, TEID for S11	Serving GW IP address and tunnel endpoint identifier for S11 interface (used by MME)
eNodeB IP@ in use	The IP address of the eNodeB currently used.
eNodeB, MME UE S1AP id	Unique identity of the UE over the S1 interface within eNodeB and within MME.
Subscribed / used UE-AMBR	The aaximum aggregated up- and downlink MBR values to be shared across all Non-GBR bearers according to the subscription of the user / currently in use.
APN Restriction	Denotes restriction on the combination of types of APN for the APN associated with this EPS bearer Context.
Subscribed Charging Characteristics	e.g. normal, prepaid, flat rate and/or hot billing.
Subscribed / used RFSP Index	Index to specific RRM configuration in the E-UTRAN received from the HSS / currently used.
URRP-MME	URRP-MME indicating that HSS has requested MME to notify regarding UE reachability at MME

Table A-2: data stored in MME (part 2)

Per PDN connection:

APN in subscribed / in use	APN received from HSS / currently used. Composed of the APN Network Identifier and the APN Operator Identifier.
IP address(es)	IPv4 and/or IPv6 address(es) used for the PDN connection.
VPLMN Address Allowed	Specifies whether UE is allowed to use this APN in HPLMN only, or additionally in VPLMN.
PDN GW IP@ in use (control plane)	IP address of the PDN GW currently used for sending control plane signalling.
Location Change Report Required	Indicates whethere it is required to communicate Cell or TAI to the PDN GW with this EPS bearer Context.
EPS subscribed QoS profile	Bearer level QoS parameter values for APN's default bearer (QCI and ARP) and AMBR.
PDN GW GRE key for uplink trafic (user plane)	GRE Key assigned by the PDN GW for the S5/S8 interface for the user plane for uplink traffic (for PMIP-based S5/S8 only).

Table A-3: data stored in MME (part 3)

For each EPS Bearer within the PDN connection:

EPS Bearer ID	Unique id of bearer for one UE accessing EPS via E-UTRAN
IP@ and TEID for S1-U	IP address and tunnel endpoint identifier of Serving GW for the S1-U interfaces.
EPS bearer QoS parameters	QCI and ARP; optionally: GBR and MBR in case of GBR bearer.
EPS Bearer Charging Characteristics	Characterizes the type of charging for the user (e.g. normal, prepaid, flat rate and/or hot billing.
Charging Id	Identifies uniquely charging records generated by SGW and PDN GW.
DL/UL TFT	Downlink/Uplink Traffic Flow Template (for PMIP-based S5/S8 only).

Table A-4: data stored in MME (part 4)

Data stored in Serving GW

Selected CN operator id	Selected core network operator identity (to support network sharing as defined in TS 23.251).
MME IP@ and TEID for S11	MME IP address and tunnel endpoint identifier for the S11 interface
SGW IP@ and TEID for S11/S4 (control plane)	Serving GW IP address and tunnel endpoint identifier for S11 Interface and the S4 Interface (control plane).
SGSN IP@ and TEID for S4 (control plane)	SGSN IP address and tunnel endpoint id for the S4 interface (Used by the Serving GW).

Table A-5: data stored in Serving GW (part 1)

For each PDN Connection:

APN in use	The APN currently used. This APN shall be composed of the APN Network Identifier and the APN Operator Identifier.
PDN GW IP@ and TEID in use (control plane)	The IP address and tunnel endpoint id of PDN GW currently used for control plane signalling.
Serving GW IP@ and TEID for S5/S8 (control plane)	Serving GW IP address and tunnel endpoint id for S5/S8 for control plane signalling.
APN Restriction	Denotes the restriction on the combination of types of APN for the APN associated with this EPS bearer Context.

Table A-6: data stored in Serving GW (part 2)

For each bearer within the PDN connection:

EPS Bearer Id	Primary identifier, uniquely identifies an EPS bearer for one UE accessing via E-UTRAN.
UL & DL TFT	Uplink and downlink Traffic Flow Template
PDN GW IP@ and TEID in use (user plane)	IP address and tunnel endpoint id of PDN GW currently used for user plane traffic.
Serving GW IP@ and TEID for S5/S8 (user plane)	Serving GW IP address and tunnel endpoint id for user plane data received from PDN GW.
Serving GW IP@ and TEID for S1-U	Serving GW IP address and tunnel endpoint id for S1-U interface (used by the eNodeB)
eNodeB IP@ and TEID for S1-U	eNodeB IP address and tunnel endpoint id for S1-U interface (used by Serving GW).
Serving GW IP@ and TEID for S12	Serving GW IP address and tunnel endpoint id for the S12 interface (used by the RNC)
RNC IP@ and TEID for S12	RNC IP address tunnel endpoint id for the S12 interface (Used by the Serving GW).
S-GW IP@ and TEID for S4 (user plane)	S-GW IP address tunnel endpoint id for the S4 interface (Used by the SGSN)
SGSN IP@ and TEID for S4 (user plane)	SGSN IP address tunnel endpoint id for the S4 interface (Used by the S-GW).
EPS Bearer QoS Profile	ARP, GBR, MBR, QCI.
Charging Id	Charging identifier, identifies charging records generated by S-GW and PDN GW.
Charging Characteristics	Characterizes the type of charging for the user (e.g. normal, prepaid, flat rate and/or hot billing)

Table A-7: data stored in Serving GW (part 3)

Abbreviations

2G, 3G, 4G	2nd, 3rd, 4th generation (of mobile networks technology)
3GPP	Third Generation Partnership Project
3GPP2	Third Generation Partnership Project 2

A

AAA	Authentication, Authorization and Accounting
ACK	Acknowledgement
AF	Application Function
AKA	Authentication and Key Agreement
AMBR	Aggregate Maximum Bit Rate
ANDSF	Access Network Discovery and Selection
APN	Access Point Name
ARP	Allocation and Retention Priority
ARQ	Automatic Repeat Request
AS	Access Stratum / Application Server
ASME	Access Security Managament Entity
AuC	Authentication Center

B

BBERF	Bearer Binding and Event Reporting Function
BCCH	Broadcast Control Channel
BCH	Broadcast Channel
BS	Base Station
BSC	Base Station Controller
BTS	Base Transceiver Station

C

CAPEX	Capital Expenditure
CBC	Cell Broadcast Center
CBE	Cell Broadcast Entity
CCCH	Common Control Channel
CcoA	Co-located Care-of-address
CDMA	Code Division Multiple Access
CK	Ciphering Key
CoA	Care-of-address
CS	Circuit Switched

CSFB	CS Fallback
CSI	Connection Set Identifier
CSG	Closed Subscriber Group

D

DAB	Digital Audio Broadcast
DB	Data base
DCH	Dedicated Channel
DDCH	Dedicated Control Channel
DHCP	Dynamic Host Configuration Protocol
DL	Downlink
DNS	Domain Name Service
DRB	Data Radio Bearer
DS	Dual Stack
DSAC	Domain Specific Access Control
DSL	Digital Subscriber Line
DSMIPv6	Dual-Stack MIPv6
DTCH	Dedicated Traffic Channel
DVB	Digital Video Broadcast

E

eHSPA	Evolved High Speed Packet Access
EDGE	Enhanced Data Rates for GSM Evolution
EIR	Equipment Identity Register
EMM	EPS Mobility Management
eNodeB	evolved Node B
EPC	Evolved Packet Core
ePDG	Evolved Packet Data Gateway
EPS	Evolved Packet System
ESM	EPS Session Management
ETSI	European Telecommunication Standards Institute
ETWS	Earthquake and Tsunami Warning System
E-UTRA(N)	Evolved UTRA(N)
EV-DO	Evolved Data Optimized

F

FA	Foreign Agent
FBC	Flow Based Charging

FDD	Frequency Division Duplex
FEC	Forward Explicit Congestion
FQDN	Fully Qualified Domain Name

G

GBR	Guaranteed Bit Rate
GERAN	GSM Enhanced Radio Access Network
GGSN	Gateway GPRS Support Node
GPRS	General Radio Packet Service
GRE	Generic Routing Encapsulation
GSM	Global System for Mobile communication
GTP	GPRS Tunneling Protocol
GUTI	Globally Unique Temporary Identity
GW	Gateway

H

HA	Home Agent
HARQ	Hybrid Automatic Repeat Request
HeNodeB	Home evolved Node B
HeNodeB-GW	Home evolved Node B Gateway
HLR	Home Location Register
HNB	Home Node B
HNB-GW	Home Node B Gateway
HoA	Home (IP) Address
hPCRF	Home PCRF
HPLMN	Home PLMN
HRPD	High Rate Packet Data
HSDPA	High Speed Downlink Packet Access
HSGW	HRPD Serving Gateway
HSPA	High Speed Packet Access
HSS	Home Subscription Server
HSUPA	High Speed Uplink Packet Access
Hz	Hertz

I

ID / id	Identifier
IEEE	Institute of Electrical and Electronic Engineers
IETF	Internet Engineering Task Force

IMEI	International Mobile Equipment Identity
IMEISV	International Mobile Equipment Identity – Software Version
IMS	Internet and Multimedia Subsystem
IMT	International Mobile Telecommunications
IP	Internet Protocol
IP-CAN	IP-Connectivity Access Network
IPMS	IP Mobility Mode Selection
IPSec	Internet Protocol Security
ISP	Internet Service Provider
ITU-R	International Telecommunication Union – Radiotelecommunications Sector
I-WLAN	Interworked WLAN

K

KASME (or K_{ASME})	Security Key at Access Security Management Entity
KHz	Kilo-Hertz
K_{RRCenc}	Security Key for RRC message encryption
K_{RRCint}	Security Key for RRC message integrity protection
K_{Upenc}	Security Key for user plane encryption

L

L1, 2, 3, 4	Layer 1, 2, 3, 4 (of OSI model)
LMA	Local Mobility Anchor
LSTI	LTE/SAE Trial Initiative
LTE	Long Term Evolution

M

MAC	Media Access Control
MAG	Mobile Access Gateway
MAP	Mobile Application Part
MBMS	Multimedia Broadcast/Multicast Service
MBR	Maximum Bit Rate
MCC	Mobile Country Code
MCH	Multicast Channel
MHz	Mega-Hertz
MIMO	Multiple Input/Output
MIPv4	Mobile IP version 4

MIPv6	Mobile IP version 6
MME	Mobility Management Entity
MMTEL	Multimedia Telephony
MNC	Mobile Network Code
MOBIKE	Mobility and Multi-homing Protocol for Internet Key Exchange
MSC	Mobile Switching Center
MTU	Maximum Transmission Unit

N

NAI	Network Access Identifier
NAP	Network Access Provider
NAS	Non-Access Stratum
NB	Node B
NGMN	Next Generation Mobile Network
NGN	Next Generation Network
NSP	Network Service Provider
NW	Network

O

OAM	Operation And Maintenance
OMC	Operation and Maintenance Center
OCS	Online Charging System
OFCS	Offline Charging System
OFDM	Orhtogonal Frequency Division Multiplexing
OMA	Open Mobile Alliance
OPEX	Operational Expenditure
OTA	Over the Air

P

PBA	Proxy Binding Acknowledge
PBCH	Physical Broadcast Channel
PBU	Proxy Bining Update
PCC	Policy and Charging Coordination
PCEF	Policy and Charging Enforcement Function
PCFICH	Physical Control Format Indicator Channel
PCH	Paging Channel
PCO	Protocol Configuration Option
PCRF	Policy and Charging Rule Function

P-CSCF	Proxy-CSCF
PDCCH	Physical Downlink Control Channel
PDCP	Packet Data Convergence Protocol
PDG	Packet Data Gateway
PDN	Packet Data Network
PDP	Packet Data Protocol
PDSCH	Physical Downlink Shared Channel
PDSN	Packet Data Serving Node
PHICH	Physical Hybrid ARQ Indicator Channel
PLMN	Public Land Mobile Network
PMCH	Physical Multicast Channel
PMIP (PMIPv6)	Proxy Mobile IP (version 6)
PRACH	Physical Random Access Channel
PS	Packet Switched
PUCCH	Physical Uplink Control Channel
PUSCH	Physical Uplink Shared Channel

Q

QAM	Quadrature Amplitude Modulation
QCI	QoS Class Index
QoS	Quality of Service

R

RAB	Radio Access Bearer
RAN	Radio Access Network
RAT	Radio Access Technology
RB	Radio Bearer
Rel	(3GPP) Release
RLC	Radio Link Control
RNC	Radio Network Controller
RoHC	Robust Header Compression
RRC	Radio Resource Control
RRM	Radio Resource Management
RTP	Real-time Transport Protocol

S

SAE	System Architecture Evolution
SBLP	Service Based Local Policy

SC	Service Continuity
SCC	Service Continuity Control
SC-FDMA	Single Carrier Frequency Division Multiple Access
S-CSCF	Serving Call Session Control Function
SDF	Service Data Flow
SDP	Session Description Protocol
SectorID	Sector Address Identifier
SGSN	Serving GPRS Support Node
SIM	Subscriber Identity Module
SIP	Session Initiation Protocol
SMS	Short Message Service
SPI	Security Parameter Index
SRB	Signaling Radio Bearer
SRVCC	Single Radio Voice Call Continuity
SS	Supplementary Service
SSAC	Service Specific Access Control

T

TA	Tracking Area
TAI	Tracking Area Identity
TAU	Tracking Area Update
TCP	Transmission Control Protocol
TDD	Time Division Duplex
TEID	Tunnel Endpoint Identifier
TFT	Traffic Flow Template
TIN	Temporary Identifier used in Next update
TISPAN	Telecoms & Internet converged Services & Protocols for Advanced Networks
TOS	Type of Service
TTL	Time to Live
TR	Technical Report
TS	Technical Specification

U

UDP	User Datagram Protocol
UE	User Equipment
UICC	Universal Integrated Circuit Card
UL	Uplink
UMTS	Universal Mobile Telecommunications System

USIM	UMTS Subscriber Identity Module
USSD	Unstructured Supplementary Service Data
UTRA(N)	Universal Terrestrial Radio Access (Network)

V

VCC	Voice Call Continuity
VLR	Visited Location Register
VoIP	Voice over IP
VoLGA	Voice over LTE Generic Access
vPCRF	Visited PCRF
VPLMN	Visited PLMN

W

W-CDMA	Wideband Code Division Multiple Access
WIMAX	Worldwide Interoperability for Microwave Access
WLAN	Wireless Local Area Network

X

xDSL	Digital Subscriber Line (any technology)

Index

F

G

H

I

J

L

M

N

O

P

Q

R

S

T

U

V

W

X

Printing and Binding: Stürtz GmbH, Würzburg